Research Notes in Mathematics 42

D Graffi

Nonlinear partial differential equations in physical problems

Pitman Advanced Publishing Program
BOSTON · LONDON · MELBOURNE

Titles in this series

1 Improperly posed boundary value problems
A Carasso and A P Stone
2 Lie algebras generated by finite dimensional ideals
I N Stewart
3 Bifurcation problems in nonlinear elasticity
R W Dickey
4 Partial differential equations in the complex domain
D L Colton
5 Quasilinear hyperbolic systems and waves
A Jeffrey
6 Solution of boundary value problems by the method of integral operators
D L Colton
7 Taylor expansions and catastrophes
T Poston and I N Stewart
8 Function theoretic methods in differential equations
R P Gilbert and R J Weinacht
9 Differential topology with a view to applications
D R J Chillingworth
10 Characteristic classes of foliations
H V Pittie
11 Stochastic integration and generalized martingales
A U Kussmaul
12 Zeta-functions: An introduction to algebraic geometry
A D Thomas
13 Explicit *a priori* inequalities with applications to boundary value problems
V G Sigillito
14 Nonlinear diffusion
W E Fitzgibbon III and H F Walker
15 Unsolved problems concerning lattice points
J Hammer
16 Edge-colourings of graphs
S Fiorini and R J Wilson
17 Nonlinear analysis and mechanics: Heriot-Watt Symposium Volume I
R J Knops
18 Actions of finite abelian groups
C Kosniowski
19 Closed graph theorems and webbed spaces
M De Wilde
20 Singular perturbation techniques applied to integro-differential equations
H Grabmüller
21 Retarded functional differential equations: A global point of view
S E A Mohammed
22 Multiparameter spectral theory in Hilbert space
B D Sleeman
23 Recent applications of generalized inverses
M Z Nashed
24 Mathematical modelling techniques
R Aris
25 Singular points of smooth mappings
C G Gibson
26 Nonlinear evolution equations solvable by the spectral transform
F Calogero
27 Nonlinear analysis and mechanics: Heriot-Watt Symposium Volume II
R J Knops
28 Constructive functional analysis
D S Bridges
29 Elongational flows: Aspects of the behaviour of model elasticoviscous fluids
C J S Petrie
30 Nonlinear analysis and mechanics: Heriot-Watt Symposium Volume III
R J Knops
31 Fractional calculus and integral transforms of generalized functions
A C McBride
32 Complex manifold techniques in theoretical physics
D E Lerner and P D Sommers
33 Hilbert's third problem: scissors congruence
C-H Sah
34 Graph theory and combinatorics
R J Wilson
35 The Tricomi equation with applications to the theory of plane transonic flow
A R Manwell
36 Abstract differential equations
S D Zaidman
37 Advances in twistor theory
L P Hughston and R S Ward
38 Operator theory and functional analysis
I Erdelyi
39 Nonlinear analysis and mechanics: Heriot-Watt Symposium Volume IV
R J Knops
40 Singular systems of differential equations
S L Campbell
41 N-dimensional crystallography
R L E Schwarzenberger
42 Nonlinear partial differential equations in physical problems
D Graffi

D Graffi

University of Bologna

Nonlinear partial differential equations in physical problems

π

Pitman Advanced Publishing Program

BOSTON · LONDON · MELBOURNE

PITMAN PUBLISHING LIMITED
39 Parker Street, London WC2B 5PB

Associated Companies
Pitman Publishing Pty Ltd., Melbourne
Pitman Publishing New Zealand Ltd., Wellington
Copp Clark Pitman, Toronto

ISBN 0 273 08474 7

Preface

This Research Note contains the lectures given in October 1978 at the Bernoulli Session of the International Centre for Mechanical Sciences, Udine, Italy.

The purpose of these lectures and the subjects covered are described in Chapter I, which represents an introduction.

I am grateful to Professor Olszak, Rector of the International Centre of Mechanical Sciences, for having invited me as a lecturer at a Bernoulli Session. I am also grateful to Professor G. Fichera, coordinator of the Session, for many stimulating discussions during its course and for his encouragement to undertake the publication of the present work.

I would also like to thank Professor M.A. Sneider for her invaluable help in the preparation of this Note.

I extend my thanks to my pupils dr. G. Pettini Boschi and dr. F. Franchi for their kind assistance in carefully reading the manuscript and checking the formulas.

Bologna University

Dario Graffi

Contents

1 Introduction

INTRODUCTORY REMARKS

1.1 Nonlinear problems were introduced into Mathematical Physics long ago; more precisely, problems were considered which can lead up to (usually partial) nonlinear differential equations. Let us restrict our discussion to three typical problems.

The equations of motion of a perfect fluid, introduced in the 18th century in the Euler formulation, are well known to be nonlinear (because of the occurrence of the so-called inertial terms). The Navier-Stokes equations for a viscous fluid, which go back to the first half of the 19th century, are also nonlinear.

The equations describing finite elastic deformations, in which the Piola-Kirchhoff tensor (introduced by Piola in 1835) plays a fundamental role, are nonlinear.

The Maxwell equations of electromagnetism are linear, but the constitutive equations are nonlinear, at least for ferromagnetic bodies, even if the hysteresis effects are neglected. This was realized in the second half of the past century, and we shall come back to this point in Chapter 2.

However, except for some isolated investigations, the results obtained earlier than the last few decades are only of a general character. The attitude then was, instead, to look for suitable approximations, in order to reduce nonlinear problems to linear ones. For example, by considering only infinitesimal strains and displacements, elastic phenomena can be represented by linear equations; this fact not only allowed important mathematical developments to be obtained, but also the construction of a theory of elasticity which in several cases is well in agreement with experiments.

In recent years, however, the most important achievements of physics and technology have made it necessary to reconsider nonlinear theory from the beginning, without too many approximations. For example, the deformation of rubber cannot be considered as infinitesimal, and the large velocities which can be reached in fluids have made it necessary to take into account nonlinear terms which were often neglected. The discovery of the laser has made possible the generation of light waves of extremely large intensity. As a result some phenomena have been discovered which can only be explained by assuming the nonlinearity of the media in which such light waves propagate, and this in turn gave rise to nonlinear optics, which will be considered in Chapter 6.

Several further examples of nonlinearity could be mentioned. On the other hand, in recent years there has been considerable mathematical progress in nonlinear (ordinary and partial) differential equations: it is, thus, conceivable that mathematics is now in a sufficiently developed stage to yield a rigorous treatment of nonlinear problems arising from physics and technology.

Hence, in these lectures, some physical problems will be recalled which can be reduced to nonlinear equations, and in this connection the not always rigorous methods employed in their treatment by mathematical physicists will be introduced. It is hoped that the readers will indicate to me whether and how such methods can be made rigorous or, more generally, what are the rigorous methods best suited to the investigation of the nonlinear problems that we will treat. We will give a more detailed treatment of the electromagnetism of rest bodies, not only because of my personal preference for this subject, but also because, in my opinion, classical electromagnetism is the simplest theory in mathematical physics from the conceptual point of view.

Furthermore, we will often speak of uniqueness theorems for the nonlinear equations that we shall meet. Such theorems have represented, as one says in Italian, "il mio pallino" (the English version of this idiomatic expression has been suggested to me by Professore Truesdell: "the bee in my bonnet") of my scientific life.

We shall, therefore, describe several methods for obtaining uniqueness theorems; however, in general, they are not coupled to existence theorems within the same function class, so that they cannot be considered fully rigorous.

Generally speaking, the classical (or, better, the old) mathematical physicists have often established, assuming the existence of solutions, uniqueness theorems, as, for example, is shown by the history of the Dirichlet problem.

Anyway, the uniqueness theorems point out the correct way of dealing with any problem in mathematical physics and, since sometimes, by chance, the solution of a nonlinear equation may be found, only a uniqueness theorem can ensure that the solution represents the actual solution of the physical problem under investigation.

Let us, now, give a short summary of the subject of the lectures that are collected together to form these Notes.

In Chapter 2, we will deal with electromagnatism. In particular, we will insist on stationary magnetic induction in ferromagnetic bodies, whose hysteresis can be neglected.

Chapter 3 is devoted to heat transfer problems, which also possess nonlinear boundary conditions.

Chapter 4 is devoted to fluid mechanics. Uniqueness theorems will be obtained both for the incompressible and the compressible case, and also for unbounded domains; we will also mention some results on the stability of fluid motion.

In Chapter 5 the electromagnetic field in a nonlinear plasma will be considered: we will indicate some cases in which a nonlinear system can be reduced to a simple nonlinear equation, this result being of some physical interest. Finally, the successive linearization method will be mentioned: the first linearization allows the introduction of some reciprocity theorems and parallels the reciprocity theorem of elastodynamics.

Finally, closing this introduction, let us list the notations and formulae frequently employed in what follows.

1.2 A point of ordinary space is indicated by x. Time is indicated by t. If $O\,y_1\,y_2\,y_3$ is an orthogonal coordinate system, the coordinates of x are y_1, y_2, y_3; $\underline{e}_1, \underline{e}_2, \underline{e}_3$ are the unit vectors along the coordinate directions.

Given a scalar function $\varphi(x) = \varphi(y_1, y_2, y_3)$, the symbol

$$\nabla\varphi = \text{grad}\,\varphi = \frac{\partial\varphi}{\partial y_1}\underline{e}_1 + \frac{\partial\varphi}{\partial y_2}\underline{e}_2 + \frac{\partial\varphi}{\partial y_3}\underline{e}_3 = \frac{\partial\varphi}{\partial y_i}\underline{e}_i \qquad (2.1)$$

stands for the gradient of φ. In the right-hand side of (2.1), the dummy index summation convention is employed. This convention will often be used in what follows.

If

$$\underline{u} = \underline{u}(x) = \underline{u}(y_1, y_2, y_3)$$

is a vector, we have

$$\underline{u} = u_1\underline{e}_1 + u_2\underline{e}_2 + u_3\underline{e}_3 = u_i\underline{e}_i \ , \tag{2.2}$$

where u_1, u_2, u_3 are the cartesian components of $\underline{u}$.

The function u or $|\underline{u}| = (u_1^2 + u_2^2 + u_3^2)^{1/2}$ is the magnitude of the vector $\underline{u}$.

The divergence of the vector $\underline{u}$ is given, as usual, by

$$\nabla . \underline{u} = \frac{\partial u_1}{\partial y_1} + \frac{\partial u_2}{\partial y_2} + \frac{\partial u_3}{\partial y_3} = \frac{\partial u_i}{\partial y_i} \tag{2.3}$$

and the curl of the vector $\underline{u}$ by

$$\nabla \times \underline{u} = \text{rot}\, \underline{u} = \text{curl}\, \underline{u} = \begin{vmatrix} \underline{e}_1 & \underline{e}_2 & \underline{e}_3 \\ \dfrac{\partial}{\partial y_1} & \dfrac{\partial}{\partial y_2} & \dfrac{\partial}{\partial y_3} \\ u_1 & u_2 & u_3 \end{vmatrix} , \tag{2.4}$$

the last term being a symbolic determinant to be expanded according to its first row elements.

Furthermore, the Laplace operator is given by

$$\nabla^2 \varphi = \nabla(\nabla \varphi) = \frac{\partial^2 \varphi}{\partial y_1^2} + \frac{\partial^2 \varphi}{\partial y_2^2} + \frac{\partial^2 \varphi}{\partial y_3^2} \ . \tag{2.5}$$

A tensor of rank two is indicated by a Greek letter such as τ, α, β, ..., and its cartesian components are indicated by τ_{ij}, α_{ij}, β_{ij}, ... (i, j = 1,2,3). The transpose of the tensor τ is denoted by $(\tau)^T$ and its components will be τ_{ji}. A tensor is said to be symmetric if $\tau = (\tau)^T$.

The relation

$$\underline{v} = \tau \underline{u} \tag{2.6}$$

defines a vector $\underline{v}$ of components v_i (i = 1,2,3) (a summation on the index j is implicitly assumed):

$$v_i = \tau_{ij} u_j \qquad (j = 1,2,3) \ . \tag{2.7}$$

The tensor of components $v_{i/j} = \frac{\partial v_i}{\partial y_j}$ is denoted by $\nabla \underline{v}$.

Let us, also, recall the following formulae:

$$\nabla \times (\nabla \varphi) = \underline{0} ; \qquad \nabla . (\nabla \times \underline{u}) = 0 ; \tag{2.8}$$

$$\nabla . (\varphi \underline{u}) = \varphi \, \nabla . \underline{u} + \nabla \varphi . \underline{u} ; \tag{2.9}$$

$$\nabla . (\underline{u} \times \underline{v}) = \nabla \times \underline{u} . \underline{v} . - \underline{u} . \nabla \times \underline{v} ; \tag{2.10}$$

$$\nabla \underline{u} . \underline{u} = \frac{1}{2} \nabla u^2 . \tag{2.11}$$

Given a domain $\mathscr{D}$ of volume S and boundary σ, under very mild assumptions on σ and $\underline{u}$, the following relation holds

$$\int_S \nabla . \underline{u} \, dS = \int_\sigma \underline{u} . \underline{n} \, d\sigma , \tag{2.12}$$

where $\underline{n}$ is the outer nomal to σ. Formula (2.12) will be referred to as <u>Green's theorem, or the divergence theorem.</u>

The meaning of some further symbols will be explained as soon as they occur.

2 Nonlinear equations in electromagnetic theory

2.1 In electromagnetic theory we deal with the simple case, in which the equations representing the fundamental laws are linear ones, while only the constitutive equations can be nonlinear. Hence we will begin with this part of mathematical physics.

2.2 As is known, in electromagnetic theory some vector fields are considered, denoted by $\underline{E}$, $\underline{H}$, $\underline{D}$, $\underline{B}$, $\underline{J}$, that are called, for short, the electromagnetic vectors, and they have the following meaning: electric field, magnetic field, electric displacement vector, magnetic induction vector and current density, respectively. These vectors are functions of the point x in space and of the time t. Hence, for the electric field vector we should write $\underline{E}(x,t)$, and analogously for the remaining fields. However, if there is no possibility of confusion, we shall simply write $\underline{E}$, $\underline{H}$, and so on.

The general laws of electricity and magnetism satisfied by the electromagnetic vectors can, in their most general form, be expressed by well known integral relations, which yield, as an immediate consequence, the Maxwell equations:

$$\nabla \times \underline{H} = \frac{\partial \underline{D}}{\partial t} + \underline{J} \ , \tag{2.1}$$

$$\nabla \times \underline{E} = -\frac{\partial \underline{B}}{\partial t} \ , \tag{2.2}$$

which are valid for any time t and any position x, with some additional conditions, as we shall see later on, on the separation surface between media of a different nature; (2.1) and (2.2) are, of course, linear equations.

We associate with (2.1) and (2.2) the equations:

$$\nabla . \underline{D} = \rho \ ; \tag{2.3}$$

$$\nabla . \underline{B} = 0 \ ; \tag{2.4}$$

$$\nabla . \underline{J} + \frac{\partial \rho}{\partial t} = 0 \ , \tag{2.5}$$

where $\rho = \rho(x,t)$ is the spatial electric density.

It is worth remarking that (2.3) can be considered as the defining relation for ρ; (2.5) is an immediate consequence of (2.1) and (2.3): as a matter of fact, taking the divergence of (2.1), we have:

$$\nabla . (\nabla \times \underline{H}) = \frac{\partial}{\partial t} \nabla . \underline{D} + \nabla . \underline{J}$$

and, since the left hand side of this equation vanishes, formula (2.5) follows on account of (2.3).

Taking the divergence of (2.2), we get:

$$\frac{\partial}{\partial t} \nabla . \underline{B} = 0 ,$$

when,(in accordance, on the other hand, with Maxwell's equations in integral form) we obtain (2.4), since it is possible to assume $\nabla . \underline{B} = 0$ for any x at a fixed time t.

To sum up, we can draw the conclusion that the essential Maxwell equations are (2.1) and (2.2). Equations (2.3), (2.4) and (2.5) are either defining equations for ρ, or are a consequence of (2.1) and (2.2).

Finally, let us remark that the general laws of electromagnetic theory in integral form yield relations valid on a bilateral surface Σ separating two media of a different nature.(1)

---

(1) Given a bilateral surface Σ and an arbitrary point $y \in \Sigma$, let the normal unit vector to Σ in y be fixed. By continuity, the normal vector is thus fixed for any other point of Σ. Consider a ball Ω of centre y, divided by Σ into two parts Ω_1 and Ω_2. By Ω_2 we shall mean the part containing $\underline{n}$. The value of the vector $\underline{u}$ at the point y, on faces (1) and (2) respectively, will be taken to be the limits (whose existence is assumed) a) $\underline{u}_1(y) = \lim_{x \to y} \underline{u}(x)$, $x \in \Omega_1$, $\underline{u}_2(y) = \lim_{x \to y} \underline{u}(x)$, $x \in \Omega_2$.

The discontinuity of $\underline{u}$ is given by $[\underline{u}] = \underline{u}_2 - \underline{u}_1$. The notation $[\underline{u} . \underline{n}] = [\underline{u}] . \underline{n}$ is self-explanatory and, furthermore, $[\underline{u}_t] = [\underline{u}] - [\underline{u} . \underline{n}] . \underline{n}$.

It may happen that, for some vector $\underline{u}$, there is a discontinuity only in the normal component. Then $[\underline{u}_t] = \underline{0}$, as happens precisely for the tangential components of $\underline{E}$ and $\underline{H}$ of (2.6) and (2.7). It may also happen that the discontinuity takes place only in the tangential component. Then $[\underline{u} . \underline{n}] = 0$, as in the case of $\underline{B}$.

If, as in the previous footnote, by $[\underline{u}_t]$ and $[\underline{u} \cdot \underline{n}]$ we denote the discontinuity across Σ of the tangential component of the vector $\underline{u}$ and of the normal one, respectively, ($\underline{n}$ being a unit vector orthogonal to Σ), we have

$$[\underline{E}_t] = \underline{0} \; ; \tag{2.6}$$

$$[\underline{H}_t] = \underline{0} \; ; \tag{2.7}$$

$$[\underline{D} \cdot \underline{n}] = \eta \; ; \tag{2.8}$$

$$[\underline{B} \cdot \underline{n}] = 0 \; ; \tag{2.9}$$

$$[\underline{J} \cdot \underline{n}] = - \frac{\partial \eta}{\partial t} \, , \tag{2.10}$$

where η is the surface electric density.

2.3 As already mentioned, the Maxwell equations can be reduced to equations (2.1) and (2.2).

Five electromagnetic vectors occur in these equations, so that we have to supply three constitutive equations among the electromagnetic vectors, that are valid for all t and x.

In ordinary electromagnetic theory these equations have the general form

$$\underline{D} = \epsilon \, \underline{E} \; ; \tag{3.1}$$

$$\underline{B} = \mu \, \underline{H} \; ; \tag{3.2}$$

$$\underline{J} = \gamma (\underline{E} + \underline{E}_i) , \tag{3.3}$$

where ϵ, μ and γ are the dielectric constant, the magnetic permeability and the conductivity, respectively, at the point x for which (3.1), (3.2) and (3.3) hold.

$\underline{E}_i$ is the so-called impressed field, which represents a schematization of the electromagnetic field source; $\underline{E}_i$ is assumed to be a known function of x and t. We remark that in isotropic media ϵ, μ and γ are scalar quantities, while in anisotropic media they are tensors of rank two (in general symmetric) and they are also usually functions at most of x.

(Continued from previous page)

The Maxwell equations hold on the opposite faces of Σ, taken, of course, as limits in the sense of a).

As can be seen, equations (3.1), (3.2) and (3.3) associated with (2.1) and (2.2), are linear: hence we can say that all the equations of ordinary electromagnetic theory are linear ones, so that we shall refer to the ordinary electromagnetic theory as the linear theory.

There are, however, examples for which at least one of the equations (3.1), (3.2) and (3.3) does not reproduce experimental data. For example, in the saturation phenomenon of ferromagnetic bodies (even if hysteresis is neglected) $\underline{B}$ is a nonlinear function of $\underline{H}$.

Hence we should write

$$\underline{B} = \underline{B}(x, \underline{H}) , \tag{3.4}$$

i.e., for any fixed x, $\underline{B}$ is a nonlinear function of $\underline{H}$; in ferromagnetic bodies (3.1) and (3.3) still hold.

In addition, there are dielectric media in which (3.2) and (3.3) are true, unlike (3.1). Then we need to write down a formula analogous to (3.4):

$$\underline{D} = \underline{D}(x, \underline{E}) , \tag{3.5}$$

i.e., for any fixed x, $\underline{D}$ is a nonlinear function of $\underline{E}$. These dielectric media (to be called, from now on, nonlinear dielectric media) occur in nonlinear optics, and they are to be studied in Chapter 6.

Finally there are nonlinear conductors, in which (3.1) and (3.2) hold, but not (3.3), which is to be replaced by

$$\underline{J} = \underline{J}(x, \underline{E} + \underline{E}_i) . \tag{3.6}$$

Anyway, a single nonlinear constitutive equation is enough to make nonlinear (and hence to complicate) the study of the electromagnetic field.

Let us remark that in what follows we will never consider nonlinear conductors, so that (3.3) will always hold. Furthermore, putting $\gamma \underline{E}_i = \underline{J}_i$, (3.3) can be rewritten as

$$\underline{J} = \gamma \underline{E} + \underline{J}_i . \tag{3.7}$$

$\underline{J}_i$ is called the impressed current.

Sometimes in what follows the electromagnetic field sources will be represented through impressed currents instead of impressed fields, and in this case $\underline{J}_i$ is taken as a known function of x and t.

2.4 Let us, now, proceed to establish some properties of the function $\underline{B}$ of $\underline{H}$ appearing in (3.4).

Since x is fixed, we shall simply write $\underline{B} = \underline{B}(\underline{H})$. We remark that the hypotheses to be stated and the results which will, thereby, be obtained hold also for the relation (3.5) between $\underline{D}$ and $\underline{E}$.

Assume, first of all, that $\underline{B}$ is an increasing function of $\underline{H}$, i.e., for $\underline{h} \neq \underline{0}$ arbitrary, we have

$$(\underline{B}(\underline{H} + \underline{h}) - \underline{B}(\underline{H})) . \underline{h} > \underline{0} . \tag{4.1}$$

Equation (4.1) is true also for a linear medium, by (3.2) because $\mu > 0$. We shall, furthermore, assume the existence of a magnetic energy, τ, whose density at the point x depends only on $\underline{H}(x)$, i.e. $\tau = \tau(\underline{H})$. According to the Poynting theorem, we assume the validity of the formula

$$d\tau = \underline{H} . d\underline{B} = \underline{H} . (\nabla_H \underline{B}) d\underline{H} , \tag{4.2}$$

where $\nabla_H \underline{B}$ is the tensor of rank two whose components are $\frac{\partial B_i}{\partial H_j}$, B_i and H_j (i, j = 1,2,3) being the cartesian components of $\underline{B}$ and $\underline{H}$ respectively.

Now (4.2) can be rewritten as

$$d\tau = H_i \frac{\partial B_i}{\partial H_j} dH_j , \tag{4.3}$$

a summation on i and j being implicitly assumed.

Hence $d\tau$ turns out to be an exact differential form in the variable dH_j, so that the following relation must hold:

$$\frac{\partial}{\partial H_j} \left(H_i \frac{\partial B_i}{\partial H_k} \right) = \frac{\partial}{\partial H_k} \left(H_i \frac{\partial B_i}{\partial H_j} \right) , \tag{4.4}$$

where the summation is now extended only to the repeated index i. Then, if δ_{ih} stands for the Kronecker symbol ($\delta_{ih} = 0, i \neq h$; $\delta_{ih} = 1, i = h$), formula

(4.4) yields

$$H_i \frac{\partial^2 B_i}{\partial H_j \partial H_k} + \frac{\partial B_i}{\partial H_k} \delta_{ij} = H_i \frac{\partial^2 B_i}{\partial H_k \partial H_j} + \frac{\partial B_i}{\partial H_j} \delta_{ik} , \tag{4.5}$$

whence

$$\frac{\partial B_j}{\partial H_k} = \frac{\partial B_k}{\partial H_j} \tag{4.6}$$

This means that the tensor $\nabla_H B$ is a symmetric one.

Furthermore, equation (4.1) implies

$$(\underline{B}(\underline{H} + \underline{h}) - \underline{B}(\underline{H})) . \underline{h} = (\nabla_H \underline{B}) \underline{h} . \underline{h} + o(h^2) > 0 ,$$

which yields

$$(\nabla_H \underline{B}) \underline{h} . \underline{h} \geq 0 . \tag{4.7}$$

Since the case in which the equality sign holds in (4.7) is excluded by assumption, we can conclude that the tensor $\nabla_H \underline{B}$ is not only symmetric, but also positive.

If the medium is isotropic, $\underline{B}$ and $\underline{H}$ have the same direction, so that we can write

$$\underline{B} = \mu(\underline{H}) \underline{H} ,$$

and always, by the isotropy, μ cannot depend on the direction of $\underline{H}$. Hence, in the isotropic case, we have:

$$\underline{B} = \mu(H^2) \underline{H} . \tag{4.8}$$

In a ferromagnetic body

$$\mu(H^2) > 0 .$$

2.5 On the basis of the preceding hypotheses, we are in a position to treat some problems concerning magnetization in ferromagnetic bodies in the stationary case, in which all electromagnetic vectors do not depend on time.

In this case (2.1) and (2.2) reduce to

$$\nabla \times \underline{H} = \underline{J} ; \tag{5.1}$$

$$\nabla \times \underline{E} = \underline{0} ; \tag{5.2}$$

$$\nabla . \underline{B} = 0 . \tag{5.3}$$

To make the problem more definite, let the domain $\mathscr{D}_1$ of ordinary space be filled up by a conductor on which $\underline{E}_i \neq \underline{0}$, and the domain $\mathscr{D}_2$ by a ferromagnetic body with $\underline{E}_i = \underline{0}$. Let the two domains be disjoint ($\mathscr{D}_1 \cap \mathscr{D}_2 = \phi$). Finally assume that, in the region external to both domains, there is vacuum (or air). As a consequence, we may assume what follows: (3.1) holds in the whole space, ϵ being the dielectric constant of the vacuum; in $\mathscr{D}_1$ both (3.2) and (3.3) hold, μ being the magnetic permeability of the vacuum; in $\mathscr{D}_2$ (3.4) and (3.3) hold with $\underline{E}_i = \underline{0}$; outside $\mathscr{D}_1 \cup \mathscr{D}_2$ we have $\gamma = 0$. Equation (5.2) then yields:

$$\underline{E} = - \nabla V , \tag{5.4}$$

where V is a function of x.

Then, by well known arguments to be omitted for brevity, $\underline{E}$ is known in the whole space, so that, by (3.3), $\underline{J}$ is known in $\mathscr{D}_1$, while $\underline{J} = \underline{0}$ outside $\mathscr{D}_1$ (in particular in $\mathscr{D}_2$).

It remains to compute $\underline{H}$, i.e. to find the magnetization of a ferromagnetic body under the action of the currents circulating in the conductor in $\mathscr{D}_1$.

Now we put

$$\underline{H}_o = \nabla \times \underline{A} , \tag{5.5}$$

where $\underline{A}$ is the so-called vector potential, i.e.

$$\underline{A}(x) = \frac{1}{4\pi} \int_S \frac{\underline{J}(y)}{|x - y|} dS , \tag{5.6}$$

S being the volume of $\mathscr{D}_1$, $y \in \mathscr{D}_1$, $|x - y|$ the distance between x and y; the integration on the right hand side is performed by keeping x fixed and letting y vary over $\mathscr{D}_1$. As is known from potential theory, $\underline{H}_o$ is continuous on the whole space and vanishes like $|x - x_o|^{-2}$, as $x \to \infty$, if the origin is taken at the point x_o. It is known that

$$\nabla \times \underline{H}_o(x) = \underline{J} , x \in \mathscr{D}_1 ; \quad \nabla \times \underline{H}_o(x) = \underline{0} , x \notin \bar{\mathscr{D}}_1 , \tag{5.7}$$

i.e. $\underline{H}_o$ is a field generated by the currents within the conductor. By means of

(3.4) we could compute $\underline{B}$, replacing $\underline{H}$ by $\underline{H}_o$. However (5.3) would be, in general, not satisfied. We have, therefore, to add a field $\underline{H}_1$ to $\underline{H}_o$, in such a way as to make possible the computation, through (3.4), of the value of $\underline{B}$ satisfying (5.3). This result is, on the other hand, an intuitive one: $\underline{H}_1$ is none other than the field due to the magnetization of the ferromagnetic body.

2.6 The problem has thus been reduced to the computation of the field $\underline{H}_1$ or, which is the same thing, to the computation of the total field

$$\underline{H} = \underline{H}_o + \underline{H}_1 . \tag{6.1}$$

Since $\underline{H}$ must satisfy (5.1) everywhere, taking into account (5.7) we have in the whole space:

$$\nabla \times \underline{H}_1 = \underline{0} , \tag{6.2}$$

whence

$$\underline{H}_1 = - \nabla W , \tag{6.3}$$

where W is the magnetic potential, which is a continuous and single valued function of the coordinates, since the tangential components of $\underline{H}$, and hence of $\underline{H}_1$, are continuous on the surface σ of the ferromagnetic body.

By analogy with $\underline{H}_o$, we shall assume $\underline{H}_1$ vanishes like $|x - x_o|^{-2}$ as $x \to \infty$.

As we shall see in the Appendix to this Chapter, it is possible to choose W in such a way that it vanishes at infinity like $|x - x_o|^{-1}$.

Then, by (3.4), we have:

$$\underline{B} = \underline{B}(\underline{H}) = \underline{B}(\underline{H}_o + \underline{H}_1) = \underline{B}(\underline{H}_o - \nabla W) \tag{6.4}$$

and, substituting into (5.3), we have the nonlinear equation to be satisfied by W within $\mathscr{D}_2$, whereas, outside $\mathscr{D}_2$, $\underline{B} = \mu \underline{H}$, $\nabla^2 W = 0$.

This equation, given $\underline{J}$ (i.e. $\underline{H}_o$), together with the boundary conditions at infinity mentioned above and with (2.9), uniquely determines W, provided that W is of class C^2 in the whole space, except for the boundary points of $\mathscr{D}_2$, where, however, (2.9) must hold.

In order to prove this theorem assume, to the contrary, the existence of two values of W, namely, W and W + v, where v must, of course, have the same properties as W. Corresponding to W we have a value of $\underline{B}$ given by (6.4), and the magnetic induction vector corresponding to W + v is:

$$\underline{B} + \underline{b} = \underline{B}(\underline{H}_o + \underline{H}_1 + \underline{h}) \,, \tag{6.5}$$

where

$$\underline{h} = - \text{grad } v \,. \tag{6.6}$$

Since (6.4) and (6.5) satisfy (5.3), we get:

$$\nabla . \underline{b} = 0 \,. \tag{6.7}$$

Furthermore, since (2.9) is also satisfied by $\underline{B} + \underline{b}$, on σ we have:

$$[\underline{b} \cdot \underline{n}] = 0 \,. \tag{6.8}$$

Given this, consider a ball Ω of radius R centred on x_o, with a boundary Σ large enough in order to contain $\mathscr{D}_2$ in its interior. After multiplication of (6.7) by v and integration over Ω, followed by some simple manipulations, taking into account (6.6) and (6.8) in order to apply Green's Theorem, we have:

$$\int_\Omega \underline{b} \cdot \underline{h} \, d\Omega + \int_\Sigma v \, \underline{b} \cdot \underline{n} \, d\Sigma = 0 \,. \tag{6.9}$$

Now let the radius R tend to infinity. Since Σ lies outside $\mathscr{D}_2$, on Σ the medium is linear and hence $\underline{b} = \mu \underline{h}$. However, on Σ, $\underline{h}$ vanishes like $\underline{H}$, i.e. like R^{-2} (of course if $x \in \Sigma$, $|x - x_o| = R$), whereas v vanishes like R^{-1}. Hence, letting R tend to infinity, the last integral on Σ tends to zero because the function under the integral sign vanishes like R^{-3}. Thus we have, recalling that $\underline{b}$ is obtained on subtracting (6.4) from (6.5) and taking into account (6.1),

$$\int_{\Omega_\infty} (\underline{B}(\underline{H} + \underline{h}) - B(\underline{H})) \,.\, \underline{h} \, d\Omega_\infty = 0 \,, \tag{6.10}$$

where Ω_∞ denotes the whole space.

Now, by virtue of (4.1), (6.10) holds only if $h \equiv 0$ for all $x \in \Omega_\infty$, hence by (6.6) $v = c$, c being some constant. But $c = 0$, because v vanishes at infinity, and hence $v = 0$ for all x, and the uniqueness of W is proved.

We remark that the theorem just proved holds even if H_o does not vanish at infinity, as happens when H_o is a uniform magnetic field.

As can be seen from the preceding proof, it is not convenient to determine W by starting from the equation satisfied by this quantity, but rather to verify that B, as given by (6.4), satisfies (5.3), (2.9) and the conditions at infinity.

2.7 Let us, now, try to solve a particular problem. This time denote by $\mathscr{D}$ the domain (of ellipsoidal shape) filled by a ferromagnetic body (supposed homogeneous), under the action of a uniform magnetic field $\underline{H}_o$ generated by currents at a large distance from $\mathscr{D}$.

To compute $\underline{H}_1$, let us introduce the vector

$$\underline{J} = \underline{B} - \mu \underline{H} , \tag{7.1}$$

i.e. $\underline{J}$ is now the vector representing the magnetic polarization (and not the current density): of course it vanishes outside $\mathscr{D}$, since there (3.2) holds with μ equal to the vacuum permeability.

Let us, now, denote by S the volume of $\mathscr{D}$ and by y a generic point in it. Set:

$$W(x) = \frac{-1}{4\pi\mu} \int_S \frac{\nabla . \underline{J}}{|x-y|} dS - \frac{1}{4\pi\mu} \int_\sigma \frac{[\underline{J} . \underline{n}]}{|y-x|} d\sigma , \tag{7.2}$$

where $\underline{n}$ is the outward normal to σ, the boundary of $\mathscr{D}$.

It can be seen at once that W is written, as is usual in magnetic theory, as the sum of a volume potential and a simple layer potential.

By well known results of potential theory, W is of class C^2, except at most on σ, and furthermore:

$$\nabla . \underline{H}_1 (x) = - \nabla^2 W(x) = - \frac{\nabla . \underline{J}(x)}{\mu} \qquad x \in \mathscr{D} , \tag{7.3}$$

$$\nabla . \underline{H}_1 (x) = - \nabla^2 W(x) = 0 \qquad x \notin \mathscr{D} . \tag{7.4}$$

In addition, on σ, because of well known properties of the simple layer potential, we have, taking into account (7.1), (6.3) and the continuity of H_o across σ,

$$[\underline{B} . \underline{n}] = \mu[(\underline{H}_o + \underline{H}_1) . \underline{n}] + [\underline{J} . \underline{n}] = \mu [\underline{H}_1 . \underline{n}] + [\underline{J} . \underline{n}] =$$

$$= - \mu [\nabla W . \underline{n}] + [\underline{J} . \underline{n}] = - [\underline{J} . \underline{n}] + [\underline{J} . \underline{n}] = 0 . \tag{7.5}$$

Hence W satisfies (2.9).

Suppose, now, $\underline{J}$ is constant in the whole of S. Hence, by (7.3), $\nabla . \underline{H}_1(x) = 0$ and, thus, $\nabla . \underline{H}(x) = 0$ for $x \in \mathscr{D}$. It then follows from (7.1) that $\nabla . \underline{B}(x) = 0$ for $x \notin \mathscr{D}$. Proceeding in the same way we can prove that $\nabla . \underline{B}(x) = 0$ for $x \in \mathscr{D}$, so that (5.3) is satisfied everywhere.

By well known properties of the Newtonian potentials, W (x) satisfies the boundary conditions as $x \to +\infty$.

It remains to determine $\underline{J}$ in such a way that, for W (x) given by (7.2), there should be an $\underline{H}_1(x)$ satisfying (6.4).

To this purpose, let us recall some properties of the potential of an ellipsoid at the points $\bar{x}$ belonging to $\mathscr{D}$. In an orthogonal coordinate system having its origin O in the centre of the ellipsoid and axes y_1, y_2, y_3 parallel to the axes of the ellipsoid, we have:

$$W = \frac{1}{\mu} (A J_1 y_1 + B J_2 y_2 + C J_3 y_3) , \tag{7.6}$$

where J_1, J_2, J_3 are the components of $\underline{J}$ along the coordinate axes, and A, B, and C are numbers depending on the shape of the ellipsoid.

We then have, denoting by $\underline{e}_1$, $\underline{e}_2$, $\underline{e}_3$ the unit vectors along the coordinate directions:

$$\underline{H}_1 = - \frac{1}{\mu} (A J_1 \underline{e}_1 + B J_2 \underline{e}_2 + C J_3 \underline{e}_3) = - \alpha \underline{J} , \tag{7.7}$$

α being a symmetric tensor of rank two, with eigenvalues $\frac{A}{\mu}$, $\frac{B}{\mu}$, $\frac{C}{\mu}$ and corresponding eigenvectors $\underline{e}_1$, $\underline{e}_2$, $\underline{e}_3$.

We thus find, by substitution into (7.1), that (6.4) is satisfied if $\underline{J}$ satisfies the equation

$$\underline{J} = \underline{B}(\underline{H}_o - \alpha \underline{J}) - \mu (\underline{H}_o - \alpha \underline{J}) . \tag{7.8}$$

We have, now, to prove the existence of a vector $\underline{J}$ satisfying (7.8). To this purpose, let us assume, according to the phenomenon of magnetic saturation, $\underline{J} = \underline{B}(\underline{H}) - \mu \underline{H}$ to be of bounded modulus for any $\underline{H}$. Given this, consider a representative space identical to the ordinary one, and let O and P be two points in this space. Consider the following vector function only of P, because O is held fixed

$$\varphi(\underline{OP}) = [\underline{B}(\underline{H}_o - \alpha \underline{OP}) - \mu(\underline{H}_o - \alpha \underline{OP})] - \underline{OP} \ . \tag{7.9}$$

As $|\underline{OP}|$ increases, since the term within square brackets is bounded, for $|\underline{OP}|$ large enough $\varphi(\underline{OP})$ tends to become parallel to $\underline{OP}$, thereby pointing in the same direction. Then, choosing a ball Ω of radius large enough, centred on O, of boundary Σ, if $P \in \Sigma$, $\varphi(\underline{OP})$ will be oriented towards the interior of Ω. By Brouwer's theorem there exists a vector $\underline{OP}_o$ such that $\varphi(\underline{OP}_o) = 0$. Hence $\underline{J}_o = \underline{OP}_o$ satisfies (7.8). Thus we have proved that the problem of determining the field generated by the magnetization of a ferromagnetic body of ellipsoidal shape subject to a uniform field is reduced to a solvable vector equation, in one and only one way, by the uniqueness theorem of the preceding Section.

Once $\underline{J}$ is known, from (7.2) we find W(x), and hence $\underline{H}$. The magnetization strength vector $\underline{J}$ is constant, and furthermore, by (7.7), the magnetic field $\underline{H}_1$ due to the magnetization is uniform within the ellipsoid. A classical theorem of linear magnetic theory has, thus, been extended to ferromagnetic bodies [1].

2.8 For the case of an isotropic sphere we have $A = B = C = \frac{1}{3}$.

Equation (7.8) reduces, then, by (7.7), (7.1) and (4.8), assuming $\underline{H}_1$ as the unknown, and indicating by μ_o the vacuum permeability, to

$$\mu(H^2)\,\underline{H} - \mu_o \underline{H} = -3\mu_o \underline{H}_1 = -3\mu_o \underline{H} + 3\mu_o \underline{H}_o \ . \tag{8.1}$$

Whence, recalling that $\underline{H}$ and $\underline{H}_o$ have a common direction and orientation (since $\mu(H^2) > 0$), we get the scalar equation:

$$\mu(H^2)\,H + 2\mu_o H = 3\mu_o H_o \ . \tag{8.2}$$

As mentioned at the end of the preceding Section, by the uniqueness theorem of Sect. 6, equation (8.2) admits a unique solution if $\mu(H^2)H$ is increasing.

However, if this condition is not satisfied, i.e. if one has, for example, (a, b and c being positive constants and $b^2 < 4ac$, since $\mu(H^2) > 0$),

$$\mu(H^2)H = aH - bH^3 + cH^5 \ , \tag{8.3}$$

by suitably choosing the constants a, b and c, equation (8.2) may admit more than one solution.

As a matter of fact, (8.2) can be rewritten as

$$(a + 2\mu_o)H - bH^3 + cH^5 = 3\mu_o H_o . \tag{8.4}$$

Putting $\psi(H)$ equal to the left hand side of (8.4), we have:

$$\psi'(H) = (a + 2\mu_o) - 3bH^2 + 5cH^4 . \tag{8.5}$$

Now, if $9b^2 > 20(a + 2\mu_o)c$ [1], equation (8.5) has two positive roots, H_1^2, H_2^2; and hence four real roots $H_1, H_2, -H_1, -H_2$ $(H_1 > 0, H_2 > 0)$.

Now, if $H_1 < H_2$, $\psi(H)$ has a maximum for $H = H_1$, since, for suitably small positive H, we have $\psi'(H) > 0$. Hence, for $H = H_2$, $\psi(H)$ has a minimum, i.e. for $H > 0$, the behaviour of $\psi(H)$ is as shown in figure 1.

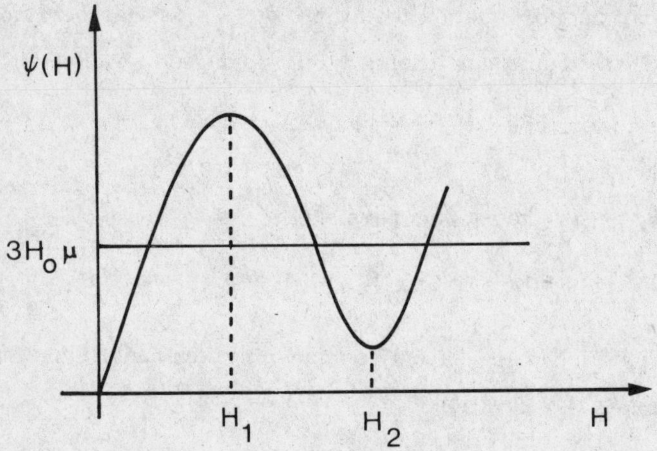

Fig. 1

It is then seen at once that, for suitable H_o, $\psi(H_1) > 3\mu H_o > \psi(H_2)$, equation (8.4) has at least three solutions, i.e. the sphere can be magnetized in three different ways. Hence, if $\underline{B}$ is a non-increasing function of $\underline{H}$, the uniqueness theorem may not hold.

2.9 Let us come back to the Maxwell equations (2.1) and (2.2), and assume, for

---

(1) By a suitable choice of a, b, c, this condition is compatible with $b^2 < 4ac$.

simplicity, $\underline{J} = \underline{0}$:

$$\nabla \times \underline{H} = \frac{\partial \underline{D}}{\partial t} \; ; \tag{9.1}$$

$$\nabla \times \underline{E} = - \frac{\partial \underline{B}}{\partial t} \; . \tag{9.2}$$

Associate with (9.1), the constitutive equations (3.4) and (3.5):

$$\underline{D} = \underline{D}(\underline{E}) \; ; \tag{9.3}$$

$$\underline{B} = \underline{B}(\underline{H}) \; . \tag{9.4}$$

Let, now, these equations be valid in a bounded domain $\mathscr{D}$, with boundary σ.

Our purpose is, now, to prove the following uniqueness theorem for (9.1), (9.2), (9.3) and (9.4):

There is at most one solution $\underline{E}(x,t)$, $\underline{H}(x,t)$ of class C^1, for all $x \in \mathscr{D}$, $t \in (0,K)$, or more briefly, $x, t \in \mathscr{D} \times (0,K)$, $K > 0$ being an arbitrary constant, and of class $C(\overline{\mathscr{D}} \times [0,K])$, provided that the initial conditions $\underline{E}(x,0)$, $\underline{H}(x,0)$ are assigned for all $x \in \mathscr{D}$, together with the boundary conditions which must specify the tangential component of $\underline{E}(x,t)$ (or $\underline{H}(x,t)$) on σ for all $t \in (0,K)$.

Since we propose, in these lectures, to describe the various methods that yield uniqueness theorems, let us apply here a method of variational type, by which M. Fabrizio [2] proved the theorem just stated under much more general conditions.

Consider two vectors $\underline{\Phi}(x,t)$ and $\underline{\Psi}(x,t)$ of class C^1 in $\overline{\mathscr{D}} \times [0,K]$, such that for any $x \in \mathscr{D}$

$$\underline{\Phi}(x,K) = \underline{0} \; ; \tag{9.5}$$

$$\underline{\Psi}(x,K) = \underline{0} \; . \tag{9.6}$$

In addition, for $x \in \sigma$, $t \in (0,K)$, let

$$\underline{\Phi}_t(x,t) = \underline{0} \; , \tag{9.7}$$

where $\underline{\Phi}_t$ is the tangential component of $\underline{\Phi}$ on σ.

After multiplication of (9.1) by $\underline{\Phi}(x,t)$ we have:

$$\nabla . (\underline{H} \times \underline{\Phi}) + \underline{H} . (\nabla \times \underline{\Phi}) = \frac{\partial}{\partial t} (\underline{\Phi} . \underline{D}) - \underline{D} . \frac{\partial \underline{\Phi}}{\partial t} . \tag{9.8}$$

Integrating first this equation on the volume S of $\mathscr{D}$, and then with respect to t on (0,K), recalling (9.5) and (9.7) we get:

$$\int_0^K dt \int_S (\underline{H} . (\nabla \times \underline{\Phi}) + \underline{D} . \frac{\partial \underline{\Phi}}{\partial t}) dS + \int_S \underline{\Phi} (x,0) . \underline{D} (x, 0) dS = 0. \tag{9.9}$$

If, on the other hand, we multiply (9.2) by $\underline{\Psi}$ (x,t), by the same procedure (suppose, to fix the ideas, the tangential component of $\underline{E}$(x,t) on σ, $t \in (0,K)$, is assigned a fixed value $\underline{g}$ (x,t)) ,we obtain:

$$\int_0^K dt \int_S (\underline{E} . (\nabla \times \underline{\Psi}) - \underline{B} . \frac{\partial \underline{\Psi}}{\partial t}) dS + \int_0^K dt \int_\sigma \underline{g} \times \underline{\Psi} . \underline{n} \, d\sigma - \int_S \underline{\Psi} (x,0) . \underline{B} (x,0) dS = 0 . \tag{9.10}$$

Now it is worth remarking that (9.1) and (9.2) are equivalent to (9.9) and (9.10), and conversely, if $\underline{E}$ and $\underline{H}$ enjoy the properties listed in the statement of the uniqueness theorem. However (9.9) and (9.10) are valid, even if $\underline{E}$ and $\underline{H}$ do not satisfy the properties just recalled and, hence, can be considered as a generalization of the Maxwell equations.

Let us, now, proceed to prove the uniqueness theorem. Suppose that (9.1) and (9.2), and hence (9.9) and (9.10), admit two distinct solutions $\underline{E}$, $\underline{H}$; $\underline{E} + \underline{e}$, $\underline{H} + \underline{h}$, with the same initial and boundary conditions. In particular we must have $\underline{e}(x,0) = \underline{0}$, $\underline{h}(x,0) = \underline{0}$, i.e. the values of $\underline{D}(x,0)$, $\underline{B}(x,0)$ are, by (9.3) and (9.4), the same for both solutions $\underline{E}$, $\underline{H}$ and $\underline{E} + \underline{e}$, $\underline{H} + \underline{h}$.

Accordingly, the tangential components of $\underline{E}$ and $\underline{E} + \underline{e}$ are both equal to $\underline{g}$ for $x \in \sigma$.

Then, substituting $\underline{E} + \underline{e}$ and $\underline{H} + \underline{h}$ into (9.9), (9.10), and taking into account (9.9) and (9.10) themselves and the initial and boundary conditions just recalled, we have:

$$\int_0^K dt \int_S (\underline{h} \cdot (\nabla \times \underline{\Phi}) + (\underline{D}(\underline{E} + \underline{e}) - \underline{D}(\underline{E})) \cdot \frac{\partial \underline{\Phi}}{\partial t}) \, dS = 0 , \tag{9.11}$$

$$\int_0^K dt \int_S (\underline{e} \cdot (\nabla \times \underline{\Psi}) - (\underline{B}(\underline{H} + \underline{h}) - \underline{B}(\underline{H})) \cdot \frac{\partial \underline{\Psi}}{\partial t}) \, dS = 0 . \tag{9.12}$$

Now we have:

$$\underline{B}(\underline{H} + \underline{h}) - \underline{B}(\underline{H}) = \int_0^1 \frac{d}{d\lambda} \underline{B}(\underline{H} + \lambda \underline{h}) \, d\lambda = \int_0^1 \nabla_H \underline{B}(\underline{H} + \lambda \underline{h}) \underline{h} \, d\lambda = \beta(x,t) \, \underline{h} , \tag{9.13}$$

where

$$\beta(x,t) = \int_0^1 \nabla_H \underline{B}(\underline{H} + \lambda \underline{h}) \, d\lambda . \tag{9.14}$$

We have written $\beta(x,t)$ since $\nabla_H(\underline{H} + \lambda \underline{h})$ depends on x and t; of course β is a symmetric tensor of rank two as also is $\nabla_H(\underline{H} + \lambda \underline{h})$.

In an analogous way we obtain:

$$\underline{D}(\underline{E} + \underline{e}) - \underline{D}(\underline{E}) = \alpha(x,t) \underline{e} , \tag{9.15}$$

$$\alpha(x,t) = \int_0^1 \nabla_E \underline{D}(\underline{E} + \lambda \underline{e}) \, d\lambda . \tag{9.16}$$

Then, substituting (9.16) and (9.15) into (9.11) and (9.12), and summing, we get:

$$\int_0^K dt \int_S (\underline{h} \cdot (\nabla \times \underline{\Phi} - \beta(x,t) \frac{\partial \underline{\Psi}}{\partial t}) + \underline{e} \cdot (\nabla \times \underline{\Psi} + \alpha(x,t) \frac{\partial \underline{\Phi}}{\partial t})) \, dS = 0 . \tag{9.17}$$

Now, choose $\underline{\Phi}$ and $\underline{\Psi}$ such as to satisfy, in addition to the conditions (9.5), (9.6) and, on σ, (9.7), also the equations

$$\nabla \times \underline{\Phi} = \beta \frac{\partial \underline{\Psi}}{\partial t} + \underline{u} , \tag{9.18}$$

$$\nabla \times \underline{\Psi} = - \alpha \frac{\partial \underline{\Phi}}{\partial t} + \underline{v} , \tag{9.19}$$

where $\underline{u}$ and $\underline{v}$ are arbitrary vectors of class C^1.

Equations (9.18) and (9.19) are linear, analogous to the Maxwell equations (if we identify $\underline{\Phi}$ with $\underline{H}$, $\underline{\Psi}$ with $\underline{E}$, provided (3.1) and (3.2) are valid with $\beta(x,t) = \epsilon$, $\alpha(x,t) = \mu$). $\underline{u}$ corresponds to an impressed electric current $\underline{J}$ and $\underline{v}$ to an impressed magnetic current, which does not exist in nature, but which is introduced in several cases for convenience.

Since α and β depend on x and t, the analogy would be with the Maxwell equations in an anisotropic non-homogeneous medium, which is, in addition, time dependent, together with conditions (9.5), (9.6) and (9.7), which can be viewed as initial and boundary conditions.

It has already been mentioned that (9.18), (9.19) are analogous, but not identical, because substituting (9.13) and (9.15) in (2.1) and (2.2), by the correspondence between $\underline{E}$, $\underline{H}$ and $\underline{\Psi}$, $\underline{\Phi}$ the first terms on the right hand side would be, rigorously speaking, $\frac{\partial}{\partial t}(\beta\Psi)$ and $-\frac{\partial}{\partial t}(\alpha\Phi)$.

Substituting (9.18) and (9.19) into (9.17) we have:

$$\int_0^K dt \int_S (\underline{h} \cdot \underline{u} + \underline{e} \cdot \underline{v})\, dS = 0 \quad , \tag{9.20}$$

whence, by the arbitrariness of $\underline{u}$ and $\underline{v}$,

$$\underline{h} \equiv \underline{0} \; , \; \underline{e} \equiv \underline{0} \qquad x,t \in \mathscr{D} \times (0,K) \; , \tag{9.21}$$

so that the uniqueness theorem is proved.

It is worth repeating that this uniqueness theorem is obtained through an existence theorem for the solutions of the linear equations (9.18) and (9.19), and such a theorem is not always known.

It may be that, in order to prove it, we have to take into account that $\underline{B} = \underline{B}(\underline{H})$ and $\underline{D} = \underline{D}(\underline{E})$ are increasing functions. This property does not appear to have been exploited so far.

2.10 <u>Appendix</u> Let us now prove the theorem applied in Sect. 2.6:

If $\nabla W(x)$ vanishes at infinity, of order 2, $W(x)$ can be chosen in such a way as to be vanishing at infinity of order 1.

To this purpose, consider a generic point x_o of the space and the half-line a) emanating from x_o and oriented as $\overline{x_o x}$. We can write $\overline{x_o x} = \rho \underline{a}$, where $\rho \geq 0$, and $\underline{a}$ is the unit vector of the half-line.

Since on a) ∇W depends only on ρ and $\nabla W \cdot d\underline{x} = \nabla W \cdot \underline{a}\, d\rho$, we have:

$$W(x) = W(x_o) + \int_{x_o}^{x} \nabla W \cdot d\underline{x} = W(x_o) + \int_{0}^{\rho} \frac{\partial W}{\partial \rho}\, d\rho \;. \tag{10.1}$$

Now, for $\rho > \rho_o$, ρ_o large enough, ∇W vanishes at infinity at least as $|x - x_o|^{-2}$, so that

$$\left| \frac{\partial W}{\partial \rho} \right| = | \nabla W \cdot \underline{a} | \leq M \rho^{-2} \;. \tag{10.2}$$

Then the integral on the right hand side of (10.1) is convergent as $\rho \to \infty$. Hence, for $\rho \to \infty$, $x \to \infty$, the left hand side of (10.1) is a finite quantity, denoted by $W_a(\infty)$ to draw attention to the fact that we have let x tend to infinity while remaining on the half-line a). Now, let us prove that W_a does not depend on a). To this end consider another half-line emanating from x_o, specified by the unit vector $\underline{b}$. We have:

$$|W_b(\infty) - W_a(\infty)| \leq \left| \lim_{\rho \to \infty} \int_C \nabla W \cdot d\underline{C} \right| , \tag{10.3}$$

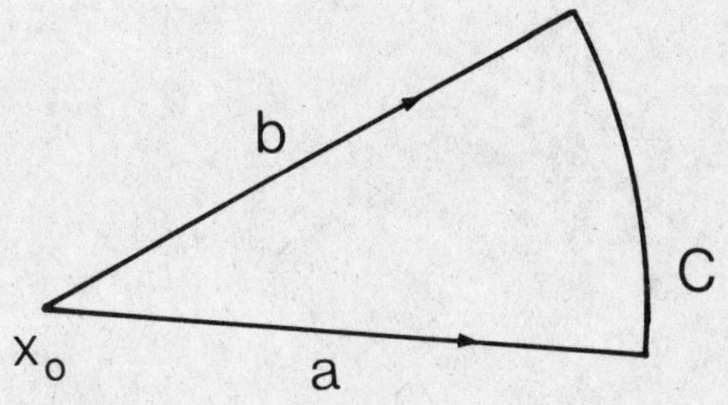

Fig. 2

where C is the arc length, between the two half-lines, of the circle (in the a),b) plane) centred on x_o and of radius ρ.

$\underline{dC}$ is an infinitesimal vector tangent to C. Then, recalling (10.2), we have:

$$\left| \int_C \nabla W \cdot \underline{dC} \right| \leq 2\pi M \rho^{-2} \rho \;.$$

Hence, as $\rho \to \infty$, the right hand side of (10.3) tends to zero, and thus $W_b(\infty) = W_a(\infty)$.

We can, thus, conclude that, more generally,

$$\lim_{x \to \infty} W(x) = W(\infty) ,$$

no matter what half-line is chosen when letting $x \to \infty$.

The result would still hold even if we do not let x tend to infinity along a half-line.

Now W (x) is determined up to a constant, so that we may choose $W(\infty) = 0$.

Denoting again by x an arbitrary point of space, and by a) the half-line parallel to $\underline{x_o x}$, we immediately obtain from (10.1)

$$W(x) - W(\infty) = - \int_{\rho}^{\infty} \frac{\partial w}{\partial \rho} \, d\rho , \tag{10.4}$$

which is the same thing as, when $\rho > \rho_o$ so that (10.2) holds,

$$W(x) \leq M \int_{\rho}^{\infty} \rho^{-2} \, d\rho = M \rho^{-1} \tag{10.5}$$

and, since $\rho = |x - x_o|$, it is proved that W (x) vanishes at infinity, of order 1.

3 Nonlinear problems in heat propagation

3.1 In heat propagation, even if we restrict our attention to isotropic media at rest, as we shall do in what follows, nonlinear problems can also be encountered. In this case, if we restrict attention to transformations which can be considered to take place at a constant volume or at a constant pressure, the nonlinearity is due to the constitutive equations or, at most, to the boundary conditions.

Let $\underline{q} = \underline{q}(x,t)$ be the heat flux vector ($x \in \mathscr{D}$, $\mathscr{D}$ is the domain occupied by the body in which heat propagates, t the time), i.e. the flux of $\underline{q}$ across a surface σ in a unit time.

As is known, the following relation holds

$$- \nabla . \underline{q} + s = c \rho \frac{\partial T}{\partial t} , \tag{1.1}$$

where $T = T(x,t)$ is the (not necessarily absolute) temperature, ρ the density, and c the specific heat, at constant volume or at constant pressure, according to the type of transformation; $s = s(x,t)$ is the so called heat source at the point x, i.e. $s(x,t)dv$ is the heat due to external origins (chemical reactions, absorption of radiation energy and so on) developing in the volume element dv, centred at x, per unit time. (1.1), of course, holds for any x and any instant t.

A constitutive equation has to be associated with (1.1), i.e. a relation between $\underline{q}$ and T, which is usually written as (Fourier's law)

$$\underline{q} = - K \operatorname{grad} T , \tag{1.2}$$

where K (usually a positive quantity) is the thermal conductivity, and K, c, ρ are assumed to depend at most on x (and never on T).

Substituting (1.2) into (1.1) we find the usual heat equation. However, experiments tell us that K as, on the other hand, c and ρ, may depend on T,

$$K = K(x,T) . \tag{1.3}$$

Substituting (1.3) into (1.2) and, then, into (1.1), we obtain a nonlinear equation in T,

$$\nabla . (K(x,T) \nabla T) + s = c\rho \frac{\partial T}{\partial t} . \tag{1.4}$$

3.2 With the equations just written down we need to associate some conditions on the boundary σ of the domain $\mathscr{D}$ occupied by the body.

Usually we divide σ in two parts, σ_1 and σ_2 ($\sigma_1 \cup \sigma_2 = \sigma$, and we may have $\sigma_1 = \sigma$). We assume P assigned on σ_1 and on σ_2 we have:

$$- K \frac{\partial T}{\partial n} = h(T - T_o) , \tag{2.1}$$

where h is the so called external thermal conductivity (T_o the external temperature), assumed to depend at most on $x \in \sigma_2$. This relation expresses the balance between the incoming and the outgoing heat on the boundary.

Sometimes, however, (2.1) does not reproduce actual phenomena. For example, if there is black body radiation heat emission from σ_2, we have to correct the right hand side of (2.1) according to Stefan's law:

$$- K \frac{\partial T}{\partial n} = \alpha (T^4 - T_o^4) , \tag{2.2}$$

where α is Stefan's constant.

Equations (2.1) and (2.2) are contained in the more general relation:

$$- K \frac{\partial T}{\partial n} = \psi(T) - \psi(T_o) , \tag{2.3}$$

where ψ is a function of T and, at most, of $x \in \sigma_2$.

(2.3) represents an example of nonlinear boundary conditions.

Assuming $\psi(T)$ to be an increasing function of T, and T to be known on σ_1 and to be given $T(x,0)$, $x \in \mathscr{D}$ (initial conditions), the nonlinear equations obtained by collecting (1.1) and (1.3) in a system as already mentioned admit, in a suitable function space, a unique solution; in the steady case (T independent of t) the initial conditions are, of course, superfluous [3]. We will not describe the proof of the theorems which can be easily obtained by the methods to be described in the next Chapter. We will, instead, see how a compatibility condition between s and $\psi(T)$

is easy to obtain when $\sigma_2 = \sigma$ and the propagation is a steady one.

Since, indeed, $\frac{\partial T}{\partial t} = 0$, equation (1.1) reduces to $\nabla . \underline{q} = -s$, which, integrated over the volume of $\mathscr{D}$, yields, taking also into account (2.3),

$$\int_S s\,dS = \int_\sigma \underline{q} . \underline{n}\, d\sigma = -\int_\sigma K \frac{\partial T}{\partial n}\, d\sigma = \int_\sigma (\psi(T) - \psi(T_o))\, d\sigma \,, \tag{2.4}$$

which is the compatibility condition we were looking for.

Essentially it describes how the heat generated by the source equals that given by the body to the external medium, as it must be, since the temperature does not depend on time.

3.3 Let us, now, proceed to the examination of a case in which, assuming K, c and ρ to be temperature-dependent, it is possible, through a suitable choice of the unknowns, to reduce (1.4) to a linear equation [4].

Let the body be homogeneous. Hence, by (1.3), K depends only on T. Set:

$$\varphi(T) = \int_a^T K(T)\,dT \,, \tag{3.1}$$

where a is a constant to be chosen in a suitable way, for example close to the presumed values of the temperature in the body.

Since, according to experiments, K is positive, φ is an increasing function of T. Hence it is possible to express T as a single-valued increasing function of φ, $T = T(\varphi)$.

Then (3.1) yields

$$\nabla \varphi = K \nabla T \,. \tag{3.2}$$

Hence

$$\nabla . (K \nabla T) = \nabla^2 \varphi \,. \tag{3.3}$$

Furthermore we have:

$$\frac{\partial \varphi}{\partial t} = K \frac{\partial T}{\partial t} \tag{3.4}$$

and, substituting into (1.4), we get:

$$\nabla^2 \varphi + s = \frac{c\rho}{K} \frac{\partial \varphi}{\partial t} . \tag{3.5}$$

Now, if $\frac{c\rho}{K}$ does not depend on temperature, (3.5) is a linear equation for φ. Anyway, if the propagation is a steady one, (3.5) becomes:

$$\nabla^2 \varphi = - s , \tag{3.6}$$

which is the well known Poisson equation, so that, if there are no sources, φ is a harmonic function.

The boundary condition (2.3) becomes:

$$- \frac{\partial \varphi}{\partial n} = \psi (T(\varphi)) - \psi (T_o) . \tag{3.7}$$

3.4 When T is steady and depends only on one coordinate, equation (1.4) can be easily studied without the introduction of φ. If, however, T depends at least on two coordinates, I believe that such an introduction is more convenient.

Consider the following simple case. Let a disk of centre O and radius b be given, which does not emit heat from its face and is not under the action of heat sources. Let the temperature be steady in the points of the disk, so that φ is a harmonic function.

Then, introducing a polar coordinate system centred on O, we have $T = T(\rho, \theta)$. Suppose we have on the boundary of the disk ($\rho = b$)

$$T(b,\theta) = T_o \sin \theta , \tag{4.1}$$

and, furthermore, assume that, near T_o, K can be represented by a linear law:

$$K = K_o + \alpha T , \tag{4.2}$$

where $\alpha > 0$. Let, in addition, $K_o - 2\alpha T_o > 0$. Then, putting $a = 0$, we have from (3.1)

$$\varphi(\rho,\theta) = K_o T + \frac{\alpha}{2} T^2 , \tag{4.3}$$

and, on the boundary of the disk, we have:

$$\varphi(b,\theta) = K_o T_o \sin\theta + \frac{\alpha}{2} T_o^2 \sin^2\theta = K_o T_o \sin\theta + \frac{\alpha}{4} T_o^2 - \frac{\alpha}{4} T_o^2 \cos 2\theta . \tag{4.4}$$

It is immediately seen that the problem of determining φ is solved by setting:

$$\varphi(\rho,\theta) = \frac{\alpha}{4} T_o^2 + K_o T_o \left(\frac{\rho}{b}\right) \sin\theta - \frac{\alpha}{4} T_o^2 \left(\frac{\rho}{b}\right)^2 \cos 2\theta . \tag{4.5}$$

φ is indeed a harmonic function, as it is the sum of three harmonic functions, and, furthermore, it satisfies (4.4).

Substituting into (4.3) we find T which is given by:

$$T = \frac{-K_o \pm \sqrt{K_o^2 + 2\alpha\varphi}}{\alpha} . \tag{4.6}$$

Now T will be real if, for any $\rho \in \mathscr{D}$,

$$K_o^2 + 2\alpha\varphi > 0 . \tag{4.7}$$

Since $\varphi(\rho,\theta)$ is harmonic, the minimum of φ will be larger than the minimum of $\varphi(b,\theta)$, which is, in turn, larger than $-K_o T_o$. Hence:

$$K_o^2 + 2\alpha\varphi > K_o^2 - 2\alpha K_o T_o \tag{4.8}$$

and, since, by assumption, $K_o > 2\alpha T_o$, the left hand side of (4.8) is positive. Hence, by (4.7), T is real. In order that T be an increasing function of φ, the minus sign of the square root in (4.6) must be excluded. Hence $T > -\frac{K_o}{\alpha}$, $K = K_o + \alpha T > K_o - K_o = 0$, i.e. K is positive for any $\rho, \theta \in \mathscr{D}$, according to our assumptions.

4 Fluid mechanics

4.1 Fluid dynamics yields, perhaps, the most interesting examples of nonlinear problems, even with linear constitutive equations. It is, indeed, well known, as already recalled in Chapter 1, that, in the so-called Euler formulation of the fluid motion (which appears often as the most convenient), the equations translating the fundamental laws of Mechanics and the conservation laws of mass and energy are nonlinear.

In order to write down these equations, consider again a fixed domain $\mathscr{D}$; let x be an arbitrary point in $\mathscr{D}$, t an arbitrary instant within the interval (0,k), $k > 0$. In some cases we may have $k = \infty$. We shall consider viscous fluids, since, from the equations for viscous fluids, it is easy to derive those for perfect ones.

Let $\underline{v}(x,t)$, $\rho(x,t)$, $p(x,t)$, $\tau(x,t)$ be the velocity, density, pressure, (symmetric) viscous stress tensor, respectively, of the particle which, at time t, lies in x. The fundamental equation of fluid motion, which is a consequence of the laws of mechanics, is given by

$$\rho \frac{d\underline{v}}{dt} = \rho \left(\frac{\partial \underline{v}}{\partial t} + (\nabla \underline{v})\underline{v}\right) = -\nabla p + \operatorname{div} \tau + \rho \underline{F} . \tag{1.1}$$

Let us try to make clear the meaning of the symbols appearing in this equation. $\frac{d\underline{v}}{dt}$, as well as $\frac{d\rho}{dt}$ to be introduced in a moment, is the material (or substantial) derivative of $\underline{v}$. To explain the other symbols, we introduce the usual orthogonal coordinate system Oy_i, $i=1,2,3$, and let v_i be the components of $\underline{v}$. Then $\nabla \underline{v}$ is the tensor of components $v_{i/j} = \frac{\partial v_i}{\partial y_j}$ $(j = 1,2,3)$; $(\nabla\underline{v})\underline{v}$ is a vector with components $v_{i/j}\, v_j$ along y_i (a sum over $j = 1,2,3$ is implicitly assumed). The components of the tensor τ are τ_{ij}, and div τ will be the vector with components. $\tau_{ij/j}$ along y_i (again a sum is assumed over j). Finally $\underline{F}$ is the force per unit mass acting on the particle at x at time t, and is, in general, a known function of ρ, t, and x. For simplicity, we assume $\underline{F}$ to be a known function only of x, and, possibly, of t. We note, also, that the nonlinear term $(\nabla v)\underline{v}$ is the so-called inertial term.

We associate with (1.1) the continuity equation

$$\frac{\partial \rho}{\partial t} + \nabla . (\rho \underline{v}) = 0 \tag{1.2}$$

or, equivalently, recalling the meaning of $\frac{d\rho}{dt} = \frac{\partial \rho}{\partial t} + \nabla \rho . \underline{v}$:

$$\frac{d\rho}{dt} + \rho \nabla . \underline{v} = 0 . \tag{1.3}$$

Finally, we need to write down the equation describing the energy conservation theorem, or, better, the first principle of Thermodynamics:

$$\rho \frac{du}{dt} = - \operatorname{div} \underline{q} + tr(\tau D) + \frac{p}{\rho} \frac{d\rho}{dt} . \tag{1.4}$$

In (1.4) u is the internal energy per unit mass, $\underline{q}$ the heat flux vector introduced in the preceding Chapter, which is related to the temperature T by (1.2) of the same Chapter, and D is the strain velocity tensor, whose cartesian components are

$$D_{ij} = \frac{1}{2} (v_{i/j} + v_{j/i}) . \tag{1.5}$$

Hence D_{ij} is a symmetric tensor.

If α is a tensor of components α_{ij}, it is known that $tr\ \alpha = \alpha_{11} + \alpha_{22} + \alpha_{33}$. Hence

$$tr\ \tau D = \tau_{ij} D_{ij} \tag{1.6}$$

(sum over i,j).

The equations written down so far are nonlinear as a consequence, as already mentioned, of mechanics as well as of thermodynamics.

Let us associate with them the additional equations, which can be called the constitutive equations beyond (1.2) of Chapter 3. i.e. (T denotes the temperature)

$$u = u(\rho, T) ; \tag{1.7}$$

$$p = p(\rho, T) \; . \tag{1.7'}$$

The functions appearing on the right-hand side are assumed to be known.

Finally, in the most usual viscous fluids (Newtonian fluids) we also assume the following relation between the terms τ and D:

$$\tau = \lambda \nabla . \underline{v} \, I + 2\eta D \; , \tag{1.8}$$

where I is the unit tensor and λ, η are the viscosity coefficients, taken to be constant for simplicity.

Furthermore, assuming everywhere positivity of the left-hand side of (1.6), which represents the energy dissipated because of viscosity, we have:

$$3\lambda + 2\eta > 0 \; , \; \eta > 0 \; . \tag{1.9}$$

Equations (1.1), (1.2), (1.4), (1.7), (1.7'), (1.8), together with (1.2) of the preceding Chapter, are $3 + 1 + 1 + 1 + 1 + 3 + 6 = 16$ scalar equations with 16 scalar unknowns (3 for $\underline{v}$, 1 for ρ, 1 for p, 1 for T, 1 for u, 3 for $\underline{q}$, 6 for τ).

For the case of a perfect fluid, by definition, there is no viscous stress, so that $\tau = 0$ and (1.8) becomes superfluous.

Before closing this Section, we remark that only (1.8) is linear; however recent investigations have led to the introduction into it of nonlinear terms. For example, if some particular fluids are concerned, we write :

$$\tau = \lambda \nabla . \underline{v} \, I + 2 \eta D + \alpha D^2 \; , \tag{1.8'}$$

α being a constant. Fluids not obeying (1.8) are called non-newtonian, and we shall not return to them.

4.2 The equations written above are very complicated, so let us review some cases in which they can be simplified, assuming the fluid to be homogeneous.

The first case arises when the fluid is incompressible. Then $\frac{d\rho}{dt} = 0$ and (1.3) yields

$$\nabla . \underline{v} = 0 \; , \tag{2.1}$$

which replaces the continuity equation.

By the homogeneity of the fluid, ρ is a positive constant, assumed to be known.

Then (1.1), (2.1) and (1.8) yield ten scalar equations for the ten unknowns (p, three components of $\underline{v}$, six components of τ) which, of course, are sufficient to determine p and $\underline{v}$ and, hence, the fluid motion.

It will be enough to study the three equations above.

The second case arises when some assumptions on (1.4) are valid: for example, when the heat absorbed by a particle, represented by $-\operatorname{div} \underline{q} + \operatorname{tr}(\tau, D)$ can be neglected, i.e. when the particle undergoes transformations which can be considered as adiabatic. Here and in similar cases, not to be recalled for sake of brevity, T can be eliminated in (1.7') so that we can write:

$$p = f(\rho) . \qquad (2.2)$$

The fluids satisfying (2.2) are called barotropic.

Of course, if we associate (2.2) with (1.1), (1.2) and (1.8), eleven scalar equations for eleven unknowns are obtained, which are thus sufficient to determine $\underline{v}, p, \rho, \tau$.

We remark that in several cases it is convenient to substitute (1.8) into (1.1), so as to eliminate τ.

Since by (1.8)

$$\tau_{ij} = \lambda \nabla . \underline{v} \, \delta_{ij} + \eta (v_{i/j} + v_{j/i}) , \qquad (2.3)$$

we have for the components of div τ along the y_i axis (recall that the index j is summed)

$$\tau_{ij/j} = \lambda \nabla . \underline{v}_{/j} \, \delta_{ij} + \eta (v_{i/jj} + v_{j/ij}) = \lambda (\nabla . \underline{v})_{/i} + \eta \nabla^2 v_i + \eta (\nabla . \underline{v})_{/i} , \qquad (2.4)$$

i.e., recalling that the components of $\nabla^2 \underline{v}$ are $\nabla^2 v_i$, and that $(\nabla . \underline{v})_{/i}$ are the components of $\nabla(\nabla . \underline{v})$, we find:

$$\operatorname{div} \tau = (\lambda + \eta) \nabla (\nabla . \underline{v}) + \eta \nabla^2 \underline{v} , \qquad (2.5)$$

whence, substituting into (1.1), we obtain the Navier-Stokes equations:

$$\rho\left(\frac{\partial \underline{v}}{\partial t} + (\nabla \underline{v})\,\underline{v}\right) = - \nabla p + (\lambda+\eta)\,\nabla(\nabla.\underline{v}) + \eta\,\nabla^2 \underline{v} + \rho\,\underline{F} \ . \tag{2.6}$$

These equations must be associated with (1.2) and, in barotropic fluids, with (2.2). When the fluid is incompressible, by (2.1), the second term on the right hand side of (2.6) vanishes.

In what follows we shall, however, prefer to consider, in place of (2.6), the system formed by (1.1), (1.2), (1.8), and (2.1) or (2.2).

In the following Sections we will proceed to establish uniqueness theorems for the equations written in this Section, under suitable initial and boundary conditions. We shall restrict ourselves to the above equations, because we are interested in establishing the method for achieving uniqueness theorems, in bounded regions as well as in unbounded ones. We remark, however, that the uniqueness theorem for the equations of Sect. 1, in a bounded domain, has been obtained by J. Serrin [5] in an important paper. Serrin's results have later been extended to an unbounded domain.

The method to be described will be called Gronwall's lemma method, because it is essentially based on this important lemma of mathematical analysis. The method is applicable only to the nonsteady case, and its difficulty consists exactly in the reduction to Gronwall's lemma. We shall, however, see that, in some cases, it is possible to achieve the uniqueness theorem also under steady conditions.

4.3 In this Section we shall state those particular cases of Gronwall's lemma in which we are interested.

Let $a(t) \geq 0$ be a function of t of class C^1. Let $a(0) = 0$ and let, in addition,

$$\frac{da}{dt} \leq Ma + \epsilon \ , \tag{3.1}$$

M and ϵ being positive constants. Then the following inequality holds:

$$a(t) \leq \epsilon\, t\, e^{Mt} \ . \tag{3.2}$$

To prove (3.2), we remark that (3.1) is equivalent to

$$e^{Mt} \frac{d}{dt} (e^{-Mt} a) \leq \epsilon . \tag{3.3}$$

Multiplying by e^{-Mt}, integrating and taking into account $a(0) = 0$, we get:

$$a(t) \leq e^{Mt} \int_0^t e^{-M\tau} \epsilon \, d\tau \tag{3.4}$$

and, since $e^{-Mt} \leq 1$, (3.2) follows immediately.

In particular, if $\epsilon = 0$, we have $a(t) = 0$ for all t, i.e. if $a(0) = 0$, $a(t) \geq 0$ and

$$\frac{da}{dt} \leq Ma , \tag{3.5}$$

we have $a(t) = 0$, for all t.

Suppose, now, $a(0) = a_o \neq 0$, $a(t) \geq 0$, $M > 0$. Let:

$$\frac{da}{dt} \leq -Ma , \tag{3.6}$$

which may also be written as

$$e^{-Mt} \frac{d}{dt} (e^{Mt} a) \leq 0 , \tag{3.7}$$

whence, dividing by e^{-Mt} and integrating, gives

$$a(t) \leq a_o e^{-Mt} . \tag{3.8}$$

This relation will also be useful in what follows. Let us, now, recall some properties of the symbol O.

Given two functions $f(x,t)$, $\varphi(x,t)$, $(x,t) \in \mathscr{D} \times (0,k)$, $\varphi(x,t) \geq 0$, we say $f(x,t) = O(\varphi(x,t))$ or, for short, $f = O(\varphi)$, if there is a positive number N such that, for any $(x,t) \in \mathscr{D} \times (0,k)$,

$$|f(x,t)| \leq N \varphi(x,t) . \tag{3.9}$$

The symbol O enjoys the following properties which are easy to prove.

If $f_1 = O(\varphi)$, $f_2 = O(\varphi)$, we have:

$$f_1 + f_2 = O(\varphi) , \tag{3.10}$$

which extends immediately to the sum of n functions $O(\varphi)$.

If $g(x,t)$ is bounded in $\mathscr{D} \times (0,k)$, we have:

$$O(gf) = O(\varphi) . \tag{3.11}$$

In particular, if $f = O(\varphi)$, $-f = O(\varphi)$, so that it is allowed in an equation to move the term $O(\varphi)$ from one side to the other without change of sign.

If $f_1 = O(\varphi_1)$, $f_2 = O(\varphi_2)$,

$$f_1 f_2 = O(\varphi_1 \varphi_2) . \tag{3.12}$$

Since, by the Cauchy inequality,

$$\varphi_1 \varphi_2 \leq \frac{1}{2} (\varphi_1^2 + \varphi_2^2) ,$$

by (3.12) we have:

$$O(f_1 f_2) = O(\varphi_1^2) + O(\varphi_2^2) . \tag{3.13}$$

Furthermore, if S is the volume of $\mathscr{D}$ and $f = O(\varphi)$,

$$\int_S f(x,t)\, dS \leq N \int_S \varphi(x,t)\, dS . \tag{3.14}$$

4.4 Let us, now, proceed to prove a uniqueness theorem for the fundamental equations of motion of an incompressible viscous fluid, namely (1.1), (2.1), (1.8), which we now rewrite (we remark that in (1.8) we take into account (2.1)):

$$\rho\left(\frac{\partial \underline{v}}{\partial t} + (\nabla \underline{v})\underline{v}\right) = -\nabla p + \operatorname{div} \tau + \rho \underline{F} , \tag{4.1}$$

$$\nabla . \underline{v} = 0 , \tag{4.2}$$

$$\tau_{ij} = \eta (v_{i/j} + v_{j/i}) . \tag{4.3}$$

More precisely, let us prove the following theorem:

Up to an arbitrary function of time for p, there is at most only one solution of (4.1), (4.2), (4 3), say $\underline{v}(x,t)$, $p(x,t)$, $\tau(x,t)$ (or, for short, $\underline{v}$, p, τ) of class C^1 in $\mathscr{D} \times (0,k)$ (and of class C in $\overline{\mathscr{D}} \times [0,k]$), provided there is given $\underline{F}$, the initial condition $\underline{v}(x,0)$ for all $x \in \mathscr{D}$, and the boundary conditions $\underline{v}(x,t)$ for any $x \in \sigma$ (boundary of $\mathscr{D}$) and for any t in $(0,k)$.

Suppose, therefore, that, in addition to the solution $\underline{v}$, p, τ of (4.1), (4.2), (4.3), there is another solution $\underline{v} + \underline{v}'$, $p+p'$, $\tau + \tau'$, satisfying the same initial and boundary conditions. Substituting $\underline{v} + \underline{v}'$, $p+p'$, $\tau + \tau'$ into (4.1), (4.2), (4.3), we obtain, recalling that $\underline{v}$, p and τ are solutions of these equations,

$$\rho \left(\frac{\partial \underline{v}'}{\partial t} + (\nabla \underline{v}) \underline{v}' + (\nabla \underline{v}') (\underline{v} + \underline{v}') \right) = - \nabla p' + \text{div } \tau' , \tag{4.4}$$

$$\nabla . \underline{v}' = 0 , \tag{4.5}$$

$$\tau'_{ij} = \eta (v'_{i/j} + v'_{j/i}) = \eta D'_{ij} ; \tag{4.6}$$

the meaning of D'_{ij} is self-explanatory.

Take the scalar product of (4.4) with $\underline{v}'$. If $(\nabla \underline{v}')^T$ denotes the tensor transpose of $\nabla \underline{v}'$, we have, on account of (4.2), (4.5) and of some well known formulae,

$$(\nabla \underline{v}') (\underline{v} + \underline{v}'). \underline{v}' = (\nabla \underline{v}')^T \underline{v}'. (\underline{v} + \underline{v}') = \nabla \frac{v'^2}{2} . (\underline{v} + \underline{v}') =$$

$$= \nabla. \left(\frac{v'^2}{2} (\underline{v} + \underline{v}') \right) - \frac{v'^2}{2} \nabla. (\underline{v} + \underline{v}') = \nabla. \frac{v'^2}{2} (\underline{v} + \underline{v}') , \tag{4.7}$$

$$\nabla p'. \underline{v}' = \nabla. (p' \underline{v}') - p' \nabla. \underline{v}' = \nabla. (p' \underline{v}') , \tag{4.8}$$

$$\text{div } \tau'. \underline{v}' = \nabla. (\tau' v') - \text{tr} (\tau' \nabla \underline{v}') . \tag{4.9}$$

Then, after multiplication by $\underline{v}'$, (4.4) becomes

$$\frac{\rho}{2} \frac{\partial}{\partial t} (v'^2) = - \rho (\nabla \underline{v}) \underline{v}'. \underline{v}' - \nabla. \left(\frac{\rho}{2} v'^2 (\underline{v} + \underline{v}') + p' \underline{v}' - \tau' \underline{v}' \right)$$

$$- \text{tr} (\tau' \nabla \underline{v}') . \tag{4.10}$$

Now, since $\underline{v}$ is of class C^1 in $\mathscr{D} \times (0,k)$, the tensor $(-\nabla \underline{v})$ will be bounded there. Hence there is a number M (given by the supremum on $\mathscr{D} \times (0,k)$ of the eigenvalues of the symmetric part of $-\nabla \underline{v}$) such that

$$-(\nabla \underline{v})\underline{v}'.\underline{v}' \leq M v'^2 \tag{4.11}$$

(we suppose $M > 0$; if this is not so, we replace M by $|M|$).

Furthermore, since τ' is symmetric as well as τ, we have:

$$\operatorname{tr}(\tau' \nabla \underline{v}') = \tau'_{ij} v'_{i/j} = \tau'_{ij} D'_{ij} = \eta (D'_{ij})^2 \geq 0 . \tag{4.12}$$

Hence from (4.10), because of (4.11) and (4.12), we get:

$$\frac{\rho}{2} \frac{\partial v'^2}{\partial t} \leq - \nabla. \left(\frac{\rho}{2} v'^2 (\underline{v} + \underline{v}') + p'\underline{v}' - \tau'\underline{v}'\right) + M \rho v'^2 , \tag{4.13}$$

and integrating over the volume S of $\mathscr{D}$:

$$\frac{d}{dt} \int_S \frac{\rho}{2} v'^2 \, dS \leq 2M \int_S \frac{\rho}{2} v'^2 \, dS - \int_\sigma \left(\frac{\rho}{2} v'^2 (\underline{v} + \underline{v}') + p'\underline{v}' - \tau'\underline{v}'\right).\underline{n} \, d\sigma . \tag{4.14}$$

Now, since $\underline{v}$ and $\underline{v} + \underline{v}'$ must satisfy the same boundary conditions, we have $\underline{v}' = 0$ on σ. Hence, putting

$$\mathscr{E}(t) = \frac{1}{2} \int_S \frac{\rho}{2} v'^2(t) \, dS , \tag{4.15}$$

(4.13) becomes

$$\frac{d\mathscr{E}}{dt} \leq 2M\mathscr{E}, \tag{4.16}$$

and, by (3.5), recalling the given initial conditions $\mathscr{E}(0) = 0$, we have for all $t \in (0,k)$,

$$\mathscr{E}(t) \equiv 0 , \tag{4.17}$$

whence $\underline{v}'(x,t) \equiv 0$, $(x,t) \in \mathscr{D} \times (0,k)$. Then, by (4.6), $\tau' = 0$ and, by (4.4), $\nabla p' = 0$, i.e. $p' = f(t)$. The uniqueness theorem is, thus, proved.

We remark that, when the fluid is perfect, the boundary conditions require only the specification of $\underline{v}.\underline{n}$, the normal component of the velocity, and, on that part σ_1 of σ where $\underline{v}.\underline{n} < 0$, i.e. from where the fluid comes, also the velocity $\underline{v}$. (σ_1

can be time dependent). In this case we have $\tau' = 0$, hence the last term in the surface integral of (4.14) vanishes.

Then, since $\underline{v}.\underline{n}$ is given, $\underline{v}'.\underline{n}$ vanishes on the whole of σ. Hence also the second term in the above integral vanishes. On σ_1 $\underline{v}$ is given, so that $\underline{v}' = \underline{0}$, and the integral of $\frac{v'^2}{2}(\underline{v} + \underline{v}').\underline{n}$ vanishes on σ_1. There remains the integral over the second part σ_2 of σ, which is always nonnegative, since on σ_2 the fluid is outgoing and, hence, $\underline{v}.\underline{n} \geq 0$. Since this integral carries a minus sign, the inequality is a fortiori true if we drop it. Hence we recover (4.9), which yields $\mathscr{E}(t) \equiv 0$ and thus the uniqueness theorem.

4.5 The results obtained in the former Section make it possible to give a short account of the stability of the motion of a viscous fluid.

The definition of the stability of the motion, taking place with velocity $\underline{v}(x,t)$, can be given in the following way. Let $\underline{v}(x,t) + \underline{v}'(x,t)$ be the velocity of a motion said to be perturbed with respect to $\underline{v}(x,t)$, $\underline{v}'(x,t)$ representing the perturbation.

The quantity $\mathscr{E}$, as defined by (4.15), is to be called energy of perturbation. Since the perturbation $\underline{v}'(x,t)$ is not supposed to be identically zero at $t = 0$, we will have $\mathscr{E}(0) \neq 0$. Alternatively, we will assume $\underline{v}'(x,t) \equiv \underline{0}$ on σ, the boundary surface of $\mathscr{D}$, i.e. motion and perturbed motion coincidence. This occurs, for example, if $\mathscr{D}$ is bounded by the surface of a solid, where the velocity of the fluid is well known to be zero in any case.

Given this, we shall say that the motion $\underline{v}(x,t)$ is stable when, for any perturbation, the corresponding energy $\mathscr{E}(t)$ tends to zero as $t \to \infty$.

To obtain a sufficient condition for stability, we first have to recall the following theorem of Serrin [6]. Denoting by d the edge of some cube containing $\mathscr{D}$ in its interior, we have:

$$\alpha d^{-2} \int_S v'^2 dS \leq \int_S \operatorname{tr} D'^2 dS, \tag{5.1}$$

where D' is defined by (4.6), α is a number equal to $\frac{3 + \sqrt{13}}{2}\pi^2 \cong 32.6$, according to Serrin's estimates; (other Authors have found a larger value for it).

Given this, consider again (4.10), and integrate it over S. Taking into account (4.11) and (4.12) we have:

$$\frac{d\mathscr{E}}{dt} \leq - \int_\sigma (\frac{\rho}{2} v'^2 (\underline{v} + \underline{v}') + p'\underline{v}' - \tau'\underline{v}') . \underline{n} \, d\sigma + 2M \int_S \frac{\rho}{2} v'^2 dS - - 2\eta \int_S \frac{D'^2_{ij}}{2} \, dS , \tag{5.2}$$

Since it has been assumed $\underline{v}' = \underline{0}$ on σ now the first integral in the right hand side of (5.2) vanishes, as proved in Sect. 4. Then, if we set

$$\beta = - (2M - \frac{2\eta\alpha}{\rho d^2}) , \tag{5.3}$$

by (5.1) we get:

$$\frac{d\mathscr{E}}{dt} \leq - \beta \mathscr{E} , \tag{5.4}$$

where, however, we have to take M to satisfy (4.12) on the whole interval $(0, \infty)$; if the unperturbed motion is a steady one, i.e. $\underline{v}$ does not depend on t, this condition is certainly satisfied.

Now let

$$\frac{\eta\alpha}{\rho d^2} > M . \tag{5.5}$$

Then we have $\beta > 0$, so that (5.4) and (3.8) yield:

$$\mathscr{E}(t) \leq \mathscr{E}(0) e^{-\beta t} , \tag{5.6}$$

i.e. (5.5) represents, according to our definition, a stability condition for the motion.

It is worth remarking that condition (5.5) ensures the uniqueness theorem for the fluid equations in the stationary case. This means that, if (5.5) holds, there is at most one solution $\underline{v}(x)$, $p(x)$, $\tau(x)$ (depending only on $x \in \mathscr{D}$) of class C^1 in $\mathscr{D}$, with a given velocity on σ.

In this case, indeed, if two solutions (depending only on x) exist, say $\underline{v}$, p, τ, $\underline{v} + \underline{v}'$, $p + p'$, $\tau + \tau'$, $\mathscr{E}$ does not depend on time and, hence, the left hand side of (5.4) vanishes. If $\mathscr{E} \neq 0$, we would get a negative number greater than zero, i.e. $\mathscr{E} = 0$, whence the uniqueness theorem follows by the same arguments of the former

Section in this Chapter.

4.6 Let us, now, proceed to prove the uniqueness theorem for barotropic fluids, i.e. for the case in which (1.1), (1.2), (1.8) and (2.2) hold.

The theorem is as follows:

There is at most one solution $\underline{v}(x,t)$, $p(x,t)$, $\rho(x,t)>0$, $\tau(x,t)$ of the above equations of class C^1 in $\mathscr{D} \times (0,k)$ (of class C in $\overline{\mathscr{D}} \times [0,k]$), provided the initial conditions $\underline{v}(x,0)$, $\rho(x,0)$ are given, together with the boundary conditions $\underline{v}(x,t)$ for any $t \in (0,k)$, $x \in \sigma$, and also ρ on that part σ_1 of σ where the fluid enters. We shall, for simplicity, assume $\underline{F} = \underline{0}$, since the case $\underline{F} \neq \underline{0}$ would add only some inessential complication.

It is worth remarking that, if one takes $f(\rho)$ to be of class C^1, by (2.2), the properties of p listed above are a consequence of those of ρ.

To prove the theorem assume, as usual, the existence in addition to $\underline{v}$, p, τ and ρ, of another solution $\underline{v} + \underline{v}'$, $p+p'$, $\tau + \tau'$, $\rho + \rho'$. Substituting $\underline{v} + \underline{v}'$, $p + p'$, $\tau + \tau'$, $\rho + \rho'$ into (1.1), (1.2), (1.8), (2.2), after some simplification we have:

$$\rho \left(\frac{\partial \underline{v}'}{\partial t} + (\nabla \underline{v}')(\underline{v} + \underline{v}') + (\nabla \underline{v})v' \right) + \rho' \frac{d}{dt}(\underline{v} + \underline{v}') = - \nabla p' + \operatorname{div} \tau' , \tag{6.1}$$

$$\frac{\partial \rho'}{\partial t} + \nabla . (\rho'(\underline{v} + \underline{v}')) + \nabla . (\rho \underline{v}') = 0 , \tag{6.2}$$

$$\tau' = \lambda I \nabla . \underline{v}' + 2\eta D' , \tag{6.3}$$

$$p' = f(\rho + \rho') - f(\rho) . \tag{6.4}$$

At this point the reader should consult the second part of Sect. 3, concerning properties of the symbol O.

Now multiply (6.1) by $\underline{v}'$. Recalling, also, (3.13), we first have:

$$\rho' \frac{d}{dt}(\underline{v} + \underline{v}').\underline{v}' = O(\rho' \underline{v}') = O\left(\frac{\rho'^2}{2}\right) + O\left(\frac{v'^2}{2}\right) , \tag{6.5}$$

and, on account of (4.7), since now $\nabla \underline{v} \neq 0$,

$$\rho(\nabla\underline{v}')(\underline{v}+\underline{v}').\underline{v}' = \rho\,\nabla.(\frac{v'^2}{2}(\underline{v}+\underline{v}')) + O(v'^2) =$$

$$= \nabla.\frac{\rho}{2}v'^2(\underline{v}+\underline{v}') - \nabla\rho.\frac{v'^2}{2}(\underline{v}+\underline{v}') + O(v'^2) = \tag{6.6}$$

$$= \nabla.(\frac{\rho}{2}v'^2(\underline{v}+\underline{v}')) + O(v'^2)\ .$$

Thus we get, recalling (4.8), (4.9) and (4.11).

$$\frac{\rho}{2}\frac{\partial v'^2}{\partial t} = -\ \nabla.(\frac{\rho}{2}v'^2(\underline{v}+\underline{v}') + p'\underline{v}' - \tau'\underline{v}') + p'\,\nabla.\underline{v}'$$
$$-\ \mathrm{tr}\,(\tau'\nabla v') + O(v'^2) + O(\rho'^2)\ . \tag{6.7}$$

Now multiply (6.2) by ρ'. Taking into account that

$$\rho'\nabla.(\rho'(\underline{v}+\underline{v}')) = \rho'\nabla\rho'.(\underline{v}+\underline{v}') + \rho'^2\,\nabla.(\underline{v}+\underline{v}') =$$
$$= \frac{1}{2}\nabla\rho'^2.(\underline{v}+\underline{v}') + \rho'^2\,\nabla.(\underline{v}+\underline{v}') = \frac{1}{2}\nabla.(\rho'^2(\underline{v}+\underline{v}')) + O(\rho'^2)\ , \tag{6.8}$$

$$\rho'\nabla.(\rho\underline{v}') = \rho'\rho\,\nabla.\underline{v}' + \rho'\nabla\rho.\underline{v}' = \rho\rho'\nabla.\underline{v}' + O(\rho'^2) + O(v'^2)\ , \tag{6.9}$$

(6.2) multiplied by ρ' becomes;

$$\frac{\partial}{\partial t}\frac{\rho'^2}{2} = -\ \nabla.(\frac{\rho'^2}{2}(\underline{v}+\underline{v}')) - \rho\rho'\nabla.\underline{v}' + O(\rho'^2) + O(v'^2)\ . \tag{6.10}$$

Furthermore (6.4) immediately yields, by the mean value theorem,

$$p' = f(\rho+\rho') - f(\rho) = O(\rho')\ . \tag{6.11}$$

Now adding (6.7) to (6.10), we remark that $\rho\,\frac{\partial v'^2}{\partial t} = \frac{\partial}{\partial t}(\rho v'^2) + O(v'^2)$, and moving $O(v'^2)$ to the right-hand side and including it in the remaining terms of order $O(v'^2)$. Then we obtain:

$$\frac{1}{2}\frac{\partial}{\partial t}(\rho v'^2 + \rho'^2) = -\ \nabla.(\frac{\rho}{2}v'^2(\underline{v}+\underline{v}') + p'\underline{v}' - \tau'\underline{v}' + \frac{\rho'^2}{2}(\underline{v}+\underline{v}'))$$
$$+\ p'\,\nabla.\underline{v}' - \rho\rho'\nabla.\underline{v}' - \mathrm{tr}(\tau'\nabla\underline{v}') + O(v'^2) + O(\rho'^2)\ . \tag{6.12}$$

The terms which are more difficult to treat are those in $\nabla.\underline{v}'$, appearing on the right-hand side of (6.12). However they can be cancelled by the terms $\mathrm{tr}\,(\tau'D')$

through a trick, due to Serrin, usually applied by people looking for uniqueness theorems.

Let us compute a lower bound for $tr\,(\tau' \nabla \underline{v}')$.

To this purpose we remark that, by (4.12),

$$tr\,(\tau' \nabla \underline{v}') = tr\,(\tau' D') . \tag{6.13}$$

Now we can decompose D' in the following well known way:

$$D' = \frac{1}{3} \nabla . \underline{v}' I + \alpha \tag{6.14}$$

and, since $tr\, D' = \nabla . \underline{v}'$, $tr\, I = 3$, we have $tr\, \alpha = 0$.

Now, substituting into (6.3), we obtain:

$$\tau' = (\lambda + \frac{2}{3} \eta)\, \nabla . \underline{v}' I + 2\eta\, \alpha . \tag{6.15}$$

Hence,

$$tr(\tau' D') = (\lambda + \frac{2}{3}\eta)\,(\nabla . \underline{v}')^2 \frac{1}{3}\, tr\, I + (\lambda + \frac{2}{3}\eta) \frac{1}{3}\, \nabla . \underline{v}' . tr(I\alpha)$$

$$+ 2\eta\, \frac{\nabla . \underline{v}'}{3} . tr(I\alpha) + 2\eta\, tr\, \alpha^2 . \tag{6.16}$$

Since $tr\,(I\alpha) = tr\, \alpha = 0$, by (6.16) we have, because of $tr(\alpha^2) \geq 0$,

$$tr(\tau' D') \geq \frac{(3\lambda + 2\eta)}{3}\, (\nabla . \underline{v}')^2 . \tag{6.17}$$

Now, let ϵ be a positive number, to be fixed at some particular value in what follows. By Cauchy's inequality, if u and v are two numbers, we have:

$$|uv| = |\frac{u}{\sqrt{\epsilon}} \sqrt{\epsilon}\, v| \leq \frac{u^2}{2\epsilon} + \frac{v^2 \epsilon}{2} . \tag{6.18}$$

Given these conditions, consider the expansion appearing on the right-hand side of (6.12), i.e.

$$p' \nabla . \underline{v}' - \rho \rho' \nabla . \underline{v}' - tr(\tau' \nabla \underline{v}') \leq \frac{p'^2}{2\epsilon} + \frac{\epsilon}{2} (\nabla . \underline{v}')^2 + \frac{\rho^2 \rho'^2}{2\epsilon}$$

$$+ \frac{\epsilon}{2} (\nabla . \underline{v}')^2 - \frac{1}{3} (3\lambda + 2\eta)\, (\nabla . \underline{v}')^2 . \tag{6.19}$$

Then, if we choose $\epsilon < \frac{1}{3}(3\lambda + 2\eta)$, by (6.11) the left-hand side of (6.19) is majorized by a term $O(\rho'^2)$.

To sum up, (6.12) can be written as

$$\frac{1}{2}\frac{\partial}{\partial t}(\rho v'^2 + \rho'^2) \le -\nabla.\left(\frac{\rho}{2} v'^2 (\underline{v} + \underline{v}') + p'\underline{v}' - \tau'\underline{v}' + \frac{\rho'^2}{2}(\underline{v} + \underline{v}')\right) + O(v'^2) + O(\rho'^2). \tag{6.20}$$

Now, since density is a positive quantity and ρ is continuous on $\overline{\mathscr{D}} \times [0,k]$, $\frac{1}{\rho}$ is bounded and, hence, $O(v'^2) = O\left(\frac{\rho v'^2}{\rho}\right) = O(\rho v'^2)$.

Let us, now, integrate (6.20) over the volume S of $\mathscr{D}$ after replacement of $O(v'^2)$ by $O(\rho v'^2)$. Applying Green's theorem to the first terms on the left-hand side, and taking into account the boundary conditions, we get:

$$\frac{1}{2}\frac{d}{dt}\int_S (\rho v'^2 + \rho'^2)\, dS \le -\frac{1}{2}\int_\sigma \rho'^2 \underline{v}.\underline{n}\, d\sigma + \int_S O(\rho v'^2 + \rho'^2)\, dS. \tag{6.21}$$

Now the integral over σ is positive by the boundary condition $\rho' = 0$, whenever $\underline{v}.\underline{n} < 0$. Then, putting

$$\mathscr{E}(t) = \frac{1}{2}\int_S (\rho v'^2 + \rho'^2)\, dS \tag{6.22}$$

and recalling the meaning of the symbol O, we have, for some $N > 0$,

$$\frac{d\mathscr{E}}{dt} \le N\mathscr{E}, \tag{6.23}$$

and, since $\mathscr{E}(0) = 0$ because, by the initial conditions, $\rho'(x,0)$ and $\underline{v}'(x,0)$ vanish, it follows that $\mathscr{E}(t) = 0$, t (0,k). Hence we have $\rho' = 0$, $\underline{v}' = \underline{0}$, $p' = 0$, $\tau' = 0$ in $\mathscr{D} \times (0,k)$. The theorem is proved.

4.7 Returning to the case of incompressible fluids, consider the case in which the domain $\mathscr{D}$ is unbounded. In this case the uniqueness theorem has been proved by myself in 1959 [7], without imposing convergence conditions at infinity for the velocity $\underline{v}$, i.e. only within the assumptions that $\underline{v}$ and $\nabla\underline{v}$ are bounded and, for $|x - x_o|$ large enough (x_o the origin),

$$|p(x,t) - p_o| \le M|x - x_o|^{-1}, \tag{7.1}$$

p_o and $M > 0$ being given numbers.

In other words, by what was described in Chapter 2, Sect. 5, the difference between p and the pressure p_o has to vanish at infinity of order 1.

More recently, this theorem has been reconsidered by Galdi and Rionero [8], [9]. by means of a much more powerful method, which I call the method of the weight function, for reasons which will become clear in a moment.

By this method they have widely generalized my results; in particular, they eliminated the hypothesis that $\nabla \underline{v}$ should be bounded, which is not always acceptable from the physical standpoint.

In order to avoid too many computations in these lectures, we will treat the weight function method only in the case of a perfect fluid under the action of no forces (which represents no restriction), in motion on the whole space S_∞, so that no boundary conditions are required.

The theorem is as follows:

There is at most one solution $\underline{v}(x,t)$, $p(x,t)$ of (1.1),(2.1) ($\tau = 0$ in (1.1)since the fluid is perfect) of class C^1 in $S_\infty \times (0,k)$ and bounded together with $\nabla \underline{v}(x,t)$ on the same domain, provided the initial condition $\underline{v}(x,0)$ is assigned on S_∞ and $p(x,t)$ tends to p_o for $x \to \infty$, of order $-\frac{1}{2} - \epsilon$, $\epsilon > 0$.

To prove this theorem, set $|x - x_o| = R(x)$, or, for short, R. Furthermore set:

$$g = e^{-\alpha R}, \tag{7.2}$$

where α is a positive constant.

g is a weight function as introduced by Galdi and Rionero, who, however, in their investigations also used different weight functions.

Now suppose that (1.1) and (2.1) (with $\tau = 0$) are satisfied by two solutions, $\underline{v}$, p, and $\underline{v} + \underline{v}'$, $p + p'$. Then (4.4) holds, and

$$\rho\left(\frac{\partial v'}{\partial t} + (\nabla \underline{v}')(\underline{v} + \underline{v}') + (\nabla \underline{v})\underline{v}'\right) = -\nabla p' . \tag{7.3}$$

Furthermore

$$\nabla . \underline{v}' = 0 . \tag{7.4}$$

Now multiply (7.3) by $g\underline{v}'$. By (4.7) we obtain

$$g\,\nabla\underline{v}'.(\underline{v}+\underline{v}').\underline{v}' = g\,\nabla\frac{v'^2}{2}.(\underline{v}+\underline{v}') = \nabla.(g\frac{v'^2}{2}(\underline{v}+\underline{v}') - \nabla g.\frac{v'^2}{2}(\underline{v}+\underline{v}') = \nabla.(g\frac{v'^2}{2}(\underline{v}+\underline{v}')) + O(g\frac{v'^2}{2}) \tag{7.5}$$

since $|\nabla g| = \alpha g$. Taking into account (4.11), we get:

$$g(\nabla\underline{v})\underline{v}'.\underline{v}' \leq M g v'^2 = O(gv'^2). \tag{7.6}$$

Furthermore we have:

$$g\,\nabla p'.\underline{v}' = g\,\nabla.(p'\underline{v}') = \nabla.(gp'\underline{v}') - \nabla g.p'\underline{v}'. \tag{7.7}$$

Now, since $|\nabla g| = \alpha g$, we get:

$$|\nabla g.p'\underline{v}'| = O(gv'^2) + O(\alpha^2 g p'^2). \tag{7.8}$$

First substituting (7.8) into (7.7) and then, (7.7), (7.6) and (7.5) into (7.3) multiplied by $g\underline{v}'$, we have:

$$\frac{\partial}{\partial t}(\frac{1}{2}\rho g v'^2) = -\nabla.(g\rho\frac{v'^2}{2}(\underline{v}+\underline{v}') + gp'\underline{v}') + O(g\rho v'^2) + O(\alpha^2 g p'^2). \tag{7.9}$$

Let us, now, integrate (7.9) over the sphere Ω of centre x_o, with radius R and surface Σ. We have:

$$\frac{d}{dt}\int_\Omega \rho g\frac{v'^2}{2}\,d\Omega = -\int_\Sigma (g\rho\frac{v'^2}{2}(\underline{v}+\underline{v}') + gp'\underline{v}').\underline{n}\,d\Sigma + \int_\Omega O(\rho g v'^2)\,d\Omega + \int_\Omega O(\alpha^2 g p'^2)\,d\Omega. \tag{7.10}$$

Letting R tend to infinity, the integral over Σ tends to zero because of the occurrence of the factor g, $\underline{v}'$ and p' being bounded. On the other hand, Ω tends to the whole space S_∞. Then, setting

$$\overline{\mathscr{E}}(t) = \frac{1}{2}\int_{S_\infty}\rho g v'^2\,dS_\infty, \tag{7.11}$$

since $O(gv'^2) = O(\rho g v'^2)$, we have:

$$\frac{d\mathscr{E}(t)}{dt} \leq M\mathscr{E}(t) + m \int_{S_\infty} \alpha^2 g p'^2 \, dS_\infty \tag{7.12}$$

(M and m are two positive constants).

Now recall that, for $|\mathbf{x}-x_o| = R > R_o$, if R_o is large enough (7.1) holds, provided -1 is replaced by $-\frac{1}{2} - \epsilon$. Thus we have, for $R > R_o$,

$$|p'| \leq M_1 R^{-\frac{1}{2}-\epsilon} \quad . \tag{7.13}$$

Now divide S_∞ in two parts, one being bounded by the sphere Ω_o of radius R_o and centre x_o, and the other, Ω_e, being the space exterior to Ω_o.

Taking also into account (7.2), set:

$$h_1(\alpha) = \alpha^2 \int_{\Omega_o} g p'^2 \, d\Omega_o \ , \tag{7.14}$$

$$h_2(\alpha) = \alpha^2 \int_{\Omega_e} g p'^2 \, d\Omega_e = 4\pi M_1 \alpha^2 \int_{R_o}^{\infty} e^{-\alpha R} R^{1-2\epsilon} \, dR \ . \tag{7.15}$$

With $\alpha R = u$, we have:

$$h_2(\alpha) = 4\pi M_1^2 \alpha^{2\epsilon} \int_{\alpha R_o}^{\infty} e^{-u} u^{1-2\epsilon} du \ .$$

Now, taking $2\epsilon < 1$ without loss of generality, since for any $\epsilon > 0$ we can always reduce the discussion to this case, we have:

$$h_2(\alpha) \leq 4\pi M_1^2 \alpha^{2\epsilon} \int_0^{\infty} u^{1-2\epsilon} e^{-u} \, du \ . \tag{7.15'}$$

Hence

$$\lim_{\alpha\to 0} h_2(\alpha) = \lim_{\alpha\to 0} h_1(\alpha) = 0. \tag{7.16}$$

Then, putting

$$h(\alpha) = h_1(\alpha) + h_2(\alpha) \ , \tag{7.17}$$

by (7.12) we get

$$\frac{d\mathscr{E}}{dt} \le M\mathscr{E} + mh(\alpha) . \tag{7.18}$$

Hence, since $\mathscr{E}(0) = 0$ by the initial conditions, by (3.2) we have:

$$\mathscr{E}(t) \le mh(\alpha)t\, e^{Mt} . \tag{7.18'}$$

Now denote by $\mathscr{E}_o$ the quantity

$$\mathscr{E}_o(t) = \frac{1}{2}\int_{\Omega_o} \rho g v'^2 d\Omega_o \ge \frac{1}{2} e^{-\alpha R_o} \int_{\Omega_o} \rho v'^2 d\Omega_o .$$

Since $\mathscr{E}_o \le \mathscr{E}(t)$, (7.18) yields:

$$e^{-\alpha R_o} \int_{\Omega_o} \frac{\rho}{2} v'^2 d\Omega_o \le mh(\alpha)te^{Mt} \le mh(\alpha)k\, e^{Mk} . \tag{7.19}$$

If we let α tend to zero, $h(\alpha) \to 0$ by (7.16) and (7.17). Thus we have:

$$\int_{\Omega_o} \frac{\rho}{2} v'^2 d\Omega_o = 0 ,$$

whence $\underline{v}'(x,t) = \underline{0}$ for all $x \in \Omega_o$, and for all $t \in (0,k)$.

Since R_o may be chosen arbitrarily large, $\underline{v}'(x,t) = \underline{0}$ for any x, and the theorem is proved.

4.8 Before closing the case of the fluid motion, let us come to the equation proposed by Burgers as a model for turbulence.

This equation is of interest here mostly as an example of a nonlinear equation, which reduces to a linear one through a well known change of unknown.

Consider the Navier equation (2.6): suppose the fluid to be incompressible, so that ρ is constant. In addition, consider a one-dimensional motion, i.e. $\underline{v}$ has only one component along an axis denoted by x. Furthermore, let $\underline{v}$ depend only on x and t. This yields:

$$\underline{v} = u(x,t)\underline{i} , \tag{8.1}$$

where $\underline{i}$ is the unit vector of the x axis.

Finally, suppose p and $\underline{F}$ to be zero. By these hypotheses, the left-hand side of (2.6) becomes

$$\rho \left[\frac{\partial u}{\partial t} + u \frac{\partial u}{\partial x} \right] \underline{i}$$

(in this case, indeed, if x coincides with the axis y_1 of Sect. 4.1, only $v_1 = u$ does not vanish, and $v_{i/j} v_j$ reduces to $u \frac{\partial u}{\partial x}$), and the right hand side becomes $\eta \frac{\partial^2 u}{\partial x^2} \underline{i}$; hence, putting $\nu = \frac{\eta}{\rho}$ (ν the kinematic viscosity coefficient) we have:

$$\frac{\partial u}{\partial t} + u \frac{\partial u}{\partial x} = \nu \frac{\partial^2 u}{\partial x^2}, \tag{8.2}$$

which is Burgers' equation.

To reduce it to a linear equation, let us put:

$$u = \frac{\partial \Psi}{\partial x}, \tag{8.3}$$

where Ψ is a function of x and t. Then (8.2) becomes

$$\frac{\partial}{\partial x} \left(\frac{\partial \Psi}{\partial t} + \frac{1}{2} \left(\frac{\partial \Psi}{\partial x}\right)^2 - \nu \frac{\partial^2 \Psi}{\partial x^2} \right) = 0 . \tag{8.4}$$

Equation (8.4), and hence (8.2), is certainly satisfied if

$$\frac{\partial \Psi}{\partial t} + \frac{1}{2} \left(\frac{\partial \Psi}{\partial x}\right)^2 = \nu \frac{\partial^2 \Psi}{\partial x^2} . \tag{8.5}$$

Now set:

$$\Psi = - 2 \nu \log \varphi . \tag{8.6}$$

We remark that ν has dimension $L^2 T^{-1}$, while (8.3) implies that Ψ has dimensions $L^2 T^{-1}$, i.e. the same as ν. Hence (8.6) is correct from the dimensional standpoint, since a nonalgebraic function such as $\log \varphi$ must be a pure number.

Now, by (8.6), we have:

$$\frac{\partial \Psi}{\partial x} = - \frac{2 \nu}{\varphi} \frac{\partial \varphi}{\partial x}, \tag{8.7}$$

$$\frac{\partial^2 \Psi}{\partial x^2} = \frac{2 \nu}{\varphi^2} \left(\frac{\partial \varphi}{\partial x}\right)^2 - \frac{2 \nu}{\varphi} \frac{\partial^2 \varphi}{\partial x^2}, \tag{8.8}$$

$$\frac{\partial \Psi}{\partial t} = - \frac{2 \nu}{\varphi} \frac{\partial \varphi}{\partial t} . \tag{8.9}$$

Substituting (8.7), (8.8), (8.9) into (8.5), after some simplification we obtain:

$$\frac{\partial \varphi}{\partial t} = \nu \frac{\partial^2 \varphi}{\partial x^2} , \tag{8.10}$$

which is the well known one-dimensional heat equation, whose properties are completely known.

It is worth remarking that, if the velocity u is known at $t = 0$, for all x $(-\infty, +\infty)$, that is if $u(x,0) = F(x)$, (8.3) yields:

$$\Psi_x(x,0) = F(x) , \tag{8.11}$$

or equivalently,

$$\Psi(x,0) = \int_0^x F(x)\, dx , \tag{8.12}$$

and hence

$$\varphi(x,0) = \exp\left\{ \frac{\int_0^x F(x)\, dx}{-2\nu} \right\}. \tag{8.13}$$

Equation (8.13) expresses the initial condition for (8.10), which, by a well known formula, gives $\varphi(x,t)$ and, hence, $u(x,t)$.

Burgers' equation can also be considered on a bounded interval, but we will not pursue this case.

5 Nonlinear plasmas. Linearization and reciprocity theorems

5.1 A further example of nonlinearity can be found in the theory of electromagnetic fields in an ionized gas, that is in a plasma.

Suppose the plasma fills a domain, which may also be unbounded. For simplicity of treatment, assume the current in the plasma to be due only to the free electron motion; let, therefore, the current density $\underline{J}$ in the plasma be $Nq\underline{v}$, where q is the electron charge, $\underline{v}$ the average velocity, and N the number of electrons per unit volume. We shall also assume, as already anticipated in Chapter 2, that possible sources of electromagnetic fields (for example, radio antennas), are represented by impressed currents $\underline{J}_i$, which are known functions of x and t. Anyway, let us always assume that the currents $\underline{J}_i$ occupy a bounded domain, even if they lie outside the domain $\mathscr{D}$ occupied by the plasma.

As in the plasma, (3.1) of Chapter 2 still holds with ϵ equal to the vacuum dielectric constant and $\underline{J} = Nq\underline{v} + \underline{J}_i$, the first Maxwell equation (2.1) of Chapter 2 becomes:

$$\nabla \times \underline{H} = \epsilon \frac{\partial \underline{E}}{\partial t} + Nq\underline{v} + \underline{J}_i \tag{1.1}$$

and, by (3.2) of Chapter 2, the second Maxwell equation becomes:

$$\nabla \times \underline{E} = -\mu \frac{\partial \underline{H}}{\partial t}, \tag{1.2}$$

where μ is the plasma magnetic permeability, in practice equal to the vacuum one.

It is worth remarking that, if the currents were due, in addition to the electrons, to n kinds of ions, of charge q_j (here and in the following, $j = 1, \ldots, n$), with mean velocity v_j and number per unit volume N_j, a term of the type $\sum_{j=1}^{n} N_j q_j v_j$ should be added to the right-hand side of (1.1).

Equations (1.1) and (1.2) contain four unknowns, $\underline{E}$, $\underline{H}$, $\underline{v}$, and N. Hence two further equations must be associated with them. The first one is obtained by applying the fundamental law of mechanics to the electrons, considered as a fluid. An equation is found analogous to (1.1) of Chapter 4:

$$Nm \frac{d\underline{v}}{dt} = Nq\,\underline{E} + Nq\mu\,\underline{v} \times (\underline{H} + \underline{H}_o) - \nu N m \underline{v} - \nabla p\,, \tag{1.3}$$

where m is the electron mass and $\frac{d\underline{v}}{dt}$ is the material derivative of the electron mean velocity; the first two terms on the right hand side correspond to the forces acting on the electrons (electric field force, magnetic field force or Lorentz force); $\underline{H}_o$ is a magnetic field generated by sources in general lying outside the plasma; it is also called the external field and is usually a given quantity; $-\nu N m \underline{v}$ is the influence on the mean electron motion due to the scattering by neutral molecules (we prove that this force is, on the average, equivalent to a force function proportional to the velocity): ν is the frequence of the collisions of the electrons with the molecules; of course $\nu > 0$.

Finally, p is the so-called pressure that an electron cluster of mean velocity $\underline{v}$ experiences from the remaining electrons. It is related to the electron number N through the relation

$$p = f(N)\,. \tag{1.4}$$

It is worth remarking first that (1.3) should be derived from the Boltzmann equation and, secondly, that, because of the occurrence of the magnetic field, p may be a tensor; however we shall take p as a scalar quantity.

Finally, the free electron number conservation yields:

$$\frac{\partial N}{\partial t} + \nabla.(N\underline{v}) = I\,, \tag{1.5}$$

where I is the electron number generated in a unit time by any ionizing source. (For example, the photoelectric effect of sunlight).

If I is negative, it has the meaning of the number of electrons which become bound, because of capture by positive ions or other similar effects.

We remark that, if (1.3) corresponds to (1.1) of Chapter 4, (1.5) corresponds to (1.2) of the same Chapter, and (1.4) to (2.2) of Chapter 4. In some sense the plasma behaves as a barotropic fluid. In any case (1.1), (1.2), (1.3), (1.4), (1.5) equal the number of unknowns, $\underline{E}$, $\underline{H}$, $\underline{v}$, p and N.

The equations just written, of course, hold for any point $x \in \mathscr{D}$ and for any time instant t.

5.2 It is worth noticing that, since the beginning, the equations (1.4), (1.5), (1.3) are nonlinear, since this is the case for $\frac{d\underline{v}}{dt}$, $f(N)$, $\operatorname{div}(N\underline{v})$, $q\underline{v} \times \underline{H}$.

It is convenient to look for some simplifying approximations, which are, more or less, usual.

First, the so called "cold plasma" is often considered, in which the electron motion is so slow (as would occur at low temperatures) that p and ∇p can be neglected in (1.3), whence (1.4) becomes superfluous. Furthermore, N is often assumed to depend on x, but not on t. This means that $\operatorname{div}(N\underline{v})$ and I in (1.5) either can be neglected or cancel each other, and the function $N(x)$ is assumed known.

Finally in (1.3) the nonlinear terms $\nabla\underline{v}.\underline{v}$ of $\frac{d\underline{v}}{dt}$ and $q.\underline{v} \times \underline{H}$ are neglected.

We, thus, obtain a linear equation:

$$m \frac{\partial \underline{v}}{\partial t} = q\,\underline{E} + q\mu\underline{v} \times \underline{H}_o - \nu m \underline{v}\ , \tag{2.1}$$

which, together with (1.1), (1.2), $N(x)$ being known, represents the linear equations of the cold plasma, largely used in applications, especially when the electromagnetic field can be considered as periodic.

We mention that, if the plasma is highly rarified, the scattering between electrons and neutral molecules is highly improbable, so that we can neglect the term $-\nu m\underline{v}$. This hypothesis is plausible even if the nonlinear terms in (1.3) are not neglected.

5.3 Consider again (1.1) (in which N is a given function of x), (1.2) and (1.3); furthermore take the case of a cold plasma, so that $\nabla p = \underline{0}$. In what follows we shall always assume, unless the contrary is explicitly stated, that $\nu = 0$, $\underline{H}_o = \underline{0}$, even if in several cases these hypotheses are unnecessary.

Since later on a method will be pointed out for studying these equations, which remain nonlinear, it is first necessary to state a uniqueness theorem.

This theorem can be of interest also because of the method of proof, which may be called the method of the contracting sphere. Actually the method is an application of the theory of characteristics; however the consideration is avoided as four-dimensional spaces are not always convenient. Anyway, let us remark that the method is very useful in dealing with problems in unbounded domains.

The theorem holds precisely when the domain $\mathscr{D}$ of the plasma coincides with the whole space, i.e. we set $\mathscr{D} = \mathscr{D}_\infty$. Let us state it in the following way:

There is at most one solution $\underline{E}(x,t), \underline{H}(x,t), \underline{v}(x,t)$ of (1.1), (1.2), (1.3) (where $p = \nu = H_o = 0$) bounded and of class C^1 on $\mathscr{D}_\infty \times (0,k)$, provided $N = N(x)$ and $J_i = J_i(x,t)$ are given for all $(x,t) \in \mathscr{D}_\infty \times (0,k)$, and provided the initial conditions $\underline{E}(x,0)$, $\underline{H}(x,0)$, $\underline{v}(x,0)$ are assigned on $\mathscr{D}_\infty$.

As usual assume the existence of two solutions of the above equations, $\underline{E}$, $\underline{H}$, $\underline{v}$ and $\underline{E} + \underline{e}$, $\underline{H} + \underline{h}$, $\underline{v} + \underline{u}$, respectively. Substituting the second solution into (1.1) and (1.2), we get:

$$\nabla \times \underline{h} = \epsilon \frac{\partial \underline{e}}{\partial t} + Nq\underline{u} \ , \tag{3.1}$$

$$\nabla \times \underline{e} = - \mu \frac{\partial \underline{h}}{\partial t} \ . \tag{3.2}$$

Substituting into (1.3), and recalling the expression of $\frac{d\underline{v}}{dt}$, we find (see the preceding Chapter, Sect. 4, with $\underline{u}$ in place of $\underline{v}'$, and Nm in place of ρ):

$$Nm\left(\frac{\partial \underline{u}}{\partial t} + (\nabla \underline{u})\underline{v} + (\nabla(\underline{v} + \underline{u}))\underline{u}\right) = Nq\underline{e} + Nq\mu\underline{u} \times (\underline{H} + \underline{h}) + \\ + Nq\mu\underline{v} \times \underline{h} \ . \tag{3.3}$$

Now multiply (3.1) and (3.2) by $\underline{e}$ and $\underline{h}$, respectively, and sum the resulting equations. We obtain:

$$\frac{\partial}{\partial t}\left(\frac{\epsilon e^2 + \mu h^2}{2}\right) = - Nq\underline{u}.\underline{e} - \nabla.(\underline{e} \times \underline{h}) \ . \tag{3.4}$$

After multiplication of (3.3) by $\underline{u}$, taking into account (4.7) of Chapter 4, we get:

$$\frac{\partial}{\partial t}\left(\frac{Nmu^2}{2}\right) = - \nabla.\left(Nm\frac{u^2}{2}\underline{v}\right) + Nm\frac{u^2}{2}\nabla.\underline{v} - Nm\nabla(\underline{u} + \underline{v})\underline{u}.\underline{u} \\ + \nabla N.m\frac{u^2}{2}\underline{v} + Nq\underline{e}.\underline{u} + Nq\mu\underline{v} \times \underline{h}.\underline{u} \ . \tag{3.5}$$

Now add (3.4) to (3.5), taking into account that $Nm\frac{u^2}{2}\nabla\underline{v} = O\left(\frac{Nmu^2}{2}\right)$,

$- \nabla N m \frac{u^2}{2} = O(\frac{N m u^2}{2})$ (assume $N > 0$ in the whole space; strictly speaking this hypothesis is not necessary), $N m \nabla(\underline{u} + \underline{v})\underline{u}.\underline{u} = O(\frac{N m u^2}{2})$, and that

$$N q \mu \underline{v} \times \underline{h}.\underline{u} = O(\mu \frac{h^2}{2}) + O(N \frac{u^2}{2}) = O(\mu \frac{h^2}{2}) + O(N m \frac{u^2}{2}) ,$$

we have:

$$\frac{\partial}{\partial t} (\frac{\epsilon e^2 + \mu h^2 + N m u^2}{2}) = - \nabla.(\underline{e} \times \underline{h} + N m \frac{u^2}{2} \underline{v}) + \\ + O(\mu \frac{h^2}{2} + N m \frac{u^2}{2}) . \tag{3.6}$$

Now consider a sphere centred on an arbitrary point x_o, and an arbitrary time instant $\bar{t} \in (0,k)$.

Consider the sphere $\Omega(t)$, whose radius $R(t)$ decreases at a velocity c equal to the maximum between the supremum of $\frac{1}{\sqrt{\epsilon \mu}}$ and the supremum of v in $\mathscr{D}_\infty \times (0,k)$ (since $\frac{1}{\sqrt{\epsilon \mu}}$ is the light speed, it is $c = \frac{1}{\sqrt{\epsilon \mu}}$).

Let R_o be the initial value of R, chosen in such a way that $R(t) = R_o - c\bar{t} = \delta$, with $\delta > 0$, i.e. the sphere still exists at time $\bar{t}$. Let, now, $\Sigma(t)$ be the surface of $\Omega(t)$. Integrate (3.6) over $\Omega(t)$. We have:

$$\frac{1}{2} \int_{\Omega(t)} \frac{\partial}{\partial t} (\epsilon e^2 + \mu h^2 + N m u^2) \, d\Omega(t) = - \int_{\Sigma(t)} (\underline{e} \times \underline{h} + N m \frac{u^2}{2} \underline{v}).\underline{n} \, d\Sigma(t) + \\ + \int_{\Omega(t)} (O(\frac{\mu h^2}{2}) + O(\frac{m N u^2}{2})) \, d\Omega(t) . \tag{3.7}$$

Now set:

$$F(x,t) = \frac{1}{2} (\epsilon e^2 + \mu h^2 + N m u^2); \tag{3.8}$$

$$\mathscr{E}(t) = \int_{\Omega(t)} F(x,t) \, d\Omega(t) . \tag{3.9}$$

By a well known theorem, we can write:

$$\frac{d}{dt} \mathscr{E}(t) = \int_{\Omega(t)} \frac{\partial F}{\partial t} \, d\Omega(t) - c \int_{\Sigma(t)} F \, d\Sigma(t) . \tag{3.10}$$

In our case we get from (3.7), recalling the meaning of the symbol O,

$$\frac{d\mathscr{E}(t)}{dt} \leq - \int_{\Sigma(t)} \frac{c}{2}(\epsilon e^2 + \mu h^2 + Nmu^2) + (\underline{e} \times \underline{h} + Nm\frac{u^2}{2}\underline{v}).\underline{n}\, d\,\Sigma(t)$$

$$+ M \int_{\Omega(t)} (\mu \frac{h^2}{2} + mN\frac{u^2}{2})\, d\,\Omega(t) . \tag{3.11}$$

Now

$$c(\frac{\epsilon}{2}e^2 + \frac{\mu}{2}h^2) + \underline{e} \times \underline{h}.\underline{n} \geq \frac{1}{\sqrt{\epsilon\mu}}(\frac{\epsilon}{2}e^2 + \frac{\mu}{2}h^2) - eh \geq$$

$$\geq \left(\sqrt[4]{\frac{\epsilon}{\mu}}\frac{e}{\sqrt{2}} - \sqrt[4]{\frac{\mu}{\epsilon}}\frac{h}{\sqrt{2}}\right)^2 \geq 0 , \tag{3.12}$$

and, since $c \geq v$, so that $c \geq \underline{v}.\underline{n}$, we have:

$$Nm\frac{u^2}{2}(c + \underline{v}.\underline{n}) \geq 0 . \tag{3.13}$$

Hence the first term on the left-hand side of (3.11) is negative: if we neglect it, the inequality is a fortiori true, and still is a fortiori true if we add $\frac{\epsilon e^2}{2}$ to the last integral. We, thus, have:

$$\frac{d\mathscr{E}(t)}{dt} \leq N\mathscr{E}(t) . \tag{3.14}$$

Then, since the initial conditions yield $e(x,0) = h(x,0) = u(x,0) = 0$ for any x, we have $\mathscr{E}(0) = 0$. Hence, by (3.5) of Chapter 4, we have $\mathscr{E}(t) \equiv 0$ for any $t \in [0,\bar{t}]$.

Hence $\underline{e}(x,t) = \underline{0}$, $\underline{h}(x,t) = \underline{0}$, $\underline{u}(x,t) = \underline{0}$ for $t \in [0,\bar{t}]$, at any point inside $\Omega(t)$, in particular in the sphere of radius δ and centre x_o.

As a consequence, we have:

$$\underline{e}(x_o,\bar{t}) = \underline{h}(x_o,\bar{t}) = \underline{u}(x_o,\bar{t}) = \underline{0} , \tag{3.15}$$

and, since x_o and $\bar{t}$ are arbitrary, the theorem is proved.

Remark. First of all, if $R_o = c\bar{t}$, (3.15) still holds. Indeed, at time $\bar{t} - \delta/c$, the sphere of radius $R_o - ct$ still exists, hence $\underline{e}(x_o, \bar{t} - \delta/c) = \underline{0}$ for any $\delta > 0$. Since δ is arbitrary and $\underline{e}(x,t)$ is continuous, $\underline{e}(x,\bar{t}) = \underline{0}$. Analogously, $\underline{h}(x,\bar{t}) = \underline{0}$ and $\underline{u}(x,\bar{t}) = \underline{0}$. Hence, by the preceding proof, it follows that $\underline{E}$, $\underline{H}$ and $\underline{u}$ are determined at x_o at time $\bar{t}$, only by the values of J_i for $t \in (0,\bar{t})$ and by the initial values of $\underline{E}$, $\underline{H}$ and $\underline{v}$ in the sphere of centre x_o and radius $R_o = c\bar{t}$. This result is intuitive: if one assumes that electromagnetic perturbations propagate at velocity c, the perturbations outside $\Omega(0)$ cannot reach x_o within a time $\bar{t}$.

Furthermore, we remark that if $\Omega(0)$ does not contain any J_i and the initial conditions are zero in $\Omega(0)$, it is immediately seen that (1.1), (1.2), (1.3) are satisfied within the sphere $\Omega(t)$ for $\underline{E}(x,t) = \underline{H}(x,t) = \underline{v}(x,t) = \underline{0}$. Then, by the uniqueness theorem, the field vanishes in $\Omega(t)$, and, in particular, at x_o, during the whole time interval $(0,\bar{t})$.

Hence, if, for $t \le 0$, there is no source of excitation, so that the initial conditions are zero everywhere, $\underline{E}$, $\underline{H}$ and $\underline{v}$ certainly vanish for $t \le \bar{t}$, at all points x_o, whose distance from the sources is $c\bar{t}$. This result is also intuitive, if electromagnetic perturbations propagate at velocity c.

5.4 Let us, now, treat a case [10], in which the nonlinear equations (1.1), (1.2), (1.3) (with $p = 0$, $\underline{H}_o = \underline{0}$) can be reduced to the linear equations of Sect. 2, together with a single nonlinear equation.

Consider again (1.1), (1.2), (2.1) with $\nu = 0$.

Assume N constant in the whole space. Denote by $\underline{E}_e$, $\underline{H}_e$, $\underline{v}_e$ some solutions of these equations.

We then have:

$$\nabla \times \underline{H}_e = \epsilon \frac{\partial \underline{E}_e}{\partial t} + N q \underline{v}_e + \underline{J}_i , \tag{4.1}$$

$$\nabla \times \underline{E}_e = - \mu \frac{\partial \underline{H}_e}{\partial t} , \tag{4.2}$$

$$m \frac{\partial \underline{v}_e}{\partial t} = q \underline{E}_e . \tag{4.3}$$

Since these equations are linear, let us call them the linear problem corresponding to the nonlinear problem to be introduced in a moment, assuming that the initial conditions satisfy the relations

$$m \nabla \times \underline{v}_e + \mu q \underline{H}_e = \underline{0} , \tag{4.4}$$

which are certainly satisfied if $\underline{v}_e$ and $\underline{H}_e$ vanish at $t = 0$.

The corresponding nonlinear problem is obtained by solving equations (1.1) (with N constant), (1.2) and (1.3), with $\nu = H_o = p = 0$. Let us make explicit the material derivative. Then:

$$\nabla \times \underline{H} = \epsilon \frac{\partial \underline{E}}{\partial t} + N q \underline{v} + \underline{J}_i , \tag{4.5}$$

$$\nabla \times \underline{E} = - \mu \frac{\partial \underline{H}}{\partial t} , \tag{4.6}$$

$$m \left(\frac{\partial \underline{v}}{\partial t} + (\nabla \underline{v}) \underline{v} \right) = q \underline{E} + \mu q \underline{v} \times \underline{H} . \tag{4.7}$$

Now, suppose the initial conditions for this problem to be the same as those of the linear problem.

To solve (4.5), (4.6), (4.7), let us put:

$$\underline{E} = \underline{E}_e + \nabla \varphi , \tag{4.8}$$

$$\underline{H} = \underline{H}_e \tag{4.9}$$

($\varphi = \varphi(x,t)$ being a function to be determined).

Inserting these into (4.5) and taking into account (4.1), we obtain:

$$\underline{0} = \epsilon \nabla \frac{\partial \varphi}{\partial t} + N q (\underline{v} - \underline{v}_e) , \tag{4.10}$$

whence

$$\underline{v} = \underline{v}_e - \frac{\epsilon \nabla \frac{\partial \varphi}{\partial t}}{N q} . \tag{4.11}$$

It is immediately verified that (4.8) and (4.9) satisfy (4.6), which reduces to (4.2).

There remains (4.7). Before verifying it, we remark that, taking the curl of (4.3), by (4.2) we have:

$$m \frac{\partial}{\partial t} (\nabla \times \underline{v}_e) = q \nabla \times \underline{E}_e = - \mu q \frac{\partial \underline{H}_e}{\partial t} ,$$

whence

$$\frac{\partial}{\partial t}\,(m\,\nabla \times \underline{v}_e + \mu\, q\, \underline{H}_e) = \underline{0}\,. \tag{4.12}$$

Hence (4.4) is true for every t. Then, for (4.9), and because for (4.11) $\nabla \times \underline{v}_e = \nabla \times \underline{v}$, we have:

$$m\,\nabla \times \underline{v} + \mu\, q\, \underline{H} = \underline{0}\,. \tag{4.13}$$

Remembering that (∇v) and $(\nabla v)^T$ are transposes, and that $(\nabla \underline{v})\underline{v} = (\nabla v)^T \underline{v} + (\nabla \times \underline{v}) \times \underline{v}$, by substitution into (4.7), we have:

$$m\left(\frac{\partial \underline{v}}{\partial t} + (\nabla v)^T \underline{v} + (\nabla \times \underline{v}) \times \underline{v}\right) = q\,\underline{E} + \mu\, q\, \underline{v} \times \underline{H} \tag{4.14}$$

and, by (4.13), the last two terms of both sides cancel each other, and, by some known manipulations, we get:

$$m\,\frac{\partial v_e}{\partial t} - \frac{m\epsilon}{Nq}\,\nabla\,\frac{\partial^2 \varphi}{\partial t^2} + m\,\nabla \frac{v^2}{2} = q\,\underline{E}_e + q\,\nabla\,\varphi\,. \tag{4.15}$$

By (4.3), this last equation reduces to

$$\nabla\left(-\frac{\partial^2 \varphi}{\partial t^2} - \frac{Nq^2}{m\epsilon}\,\varphi + \frac{Nq}{\epsilon}\, v^2\right) = \underline{0}\,, \tag{4.16}$$

which is satisfied if

$$\frac{\partial^2 \varphi}{\partial t^2} + \frac{Nq^2}{m\epsilon}\,\varphi - \frac{Nq}{\epsilon}\left(\underline{v}_e - \frac{\epsilon}{Nq}\,\nabla\,\frac{\partial \varphi}{\partial t}\right)^2 = 0\,. \tag{4.17}$$

Hence, if φ satisfies the nonlinear equation (4.17), equations (4.5), (4.6), (4.7) are satisfied by (4.8), (4.9), (4.11). It remains for us to verify the initial conditions. Since they are the same for both problems, we must have $\nabla\varphi(x,0) = \underline{0}$, because $\underline{E}(x,0)$ coincides with $\underline{E}_e(x,0)$, and $\nabla \frac{\partial \varphi}{\partial t}(x,0) = \underline{0}$, because $\underline{v}$ coincides with $\underline{v}_e(x,0)$.

These conditions are satisfied by setting

$$\varphi(x,0) = \frac{\partial \varphi}{\partial t}(x,0) = 0\,, \tag{4.18}$$

which have to be associated with (4.17).

Hence, when φ is determined through (4.17), (4.18), by the uniqueness theorem of the preceding Section, (4.8), (4.9), (4.11) satisfy all the conditions determining the nonlinear problem. In other words, the nonlinear problem is reduced to the corresponding linear problem (easier to solve, generally speaking) and to the nonlinear equation (4.17).

This last equation becomes linear if the term $(\frac{\epsilon}{Nq} \nabla \frac{\partial \varphi}{\partial t})^2$ is neglected.

The result of the present Section may not only be of some mathematical interest, but also of some physical applicability, even if some assumptions may look restrictive, as, for example, the time-independence of N. We shall come back to this point later on. However, in the proof (see the step from (4.14) to (4.15)) it is shown that the nonlinear terms of the material derivative are cancelled by terms due to the wave magnetic field. That is, the two terms are of the same order of magnitude.

We are not allowed to neglect any one of these terms without neglecting the other.

Let us also remark that the method could be applied taking the source, and hence the field, to be periodic, and thus avoiding the initial conditions.

However, to my knowledge, a uniqueness theorem still does not exist in this case.

5.5 Consider again equations (1.1), (1.2), (1.3) (where, for simplicity, we take $\underline{H}_o = \underline{0}$, $\nu = 0$), (1.4) and (1.5) (where we take $I = 0$):

$$\nabla \times \underline{H} = \epsilon \frac{\partial \underline{E}}{\partial t} + N q \underline{v} + \underline{J}_i , \tag{5.1}$$

$$\nabla \times \underline{E} = - \mu \frac{\partial \underline{H}}{\partial t} , \tag{5.2}$$

$$N m \left(\frac{\partial \underline{v}}{\partial t} + (\nabla \underline{v}) \underline{v} \right) = N q \underline{E} + N q \mu \underline{v} \times \underline{H} - \nabla p , \tag{5.3}$$

$$p = f(N) \tag{5.4}$$

$$\frac{\partial N}{\partial t} + \nabla . (N \underline{v}) = 0 . \tag{5.5}$$

We can see now that p is also taken into account; that is we are now considering the so-called "hot" plasma.

Unless the contrary is explicitly stated, equations (5.1) through (5.5) are supposed to be valid in the whole space.

Assume, now, the field sources to be proportional to some parameter η, so that

$$\underline{J}_i = \eta \, \underline{J}_{io} . \tag{5.6}$$

If $\eta = 0$, so that there are no sources, the gas may be taken to be ionized; that is $N = N_o(x)$.

Then, by (5.4), we have $p \neq 0$,

$$p = p_o = f(N_o) . \tag{5.7}$$

Then equations (5.1) through (5.3) are satisfied by putting:

$$\underline{v} = \underline{0} , \quad \underline{E}_o = \frac{\nabla p_o}{N_o q} , \quad \underline{H} = \underline{0} . \tag{5.8}$$

This is a stationary solution of (5.5).

Suppose, now, $\eta \neq 0$. To solve the above equations, let us introduce a method which, although not rigorous, is widely used in mathematical physics and allows the reduction of the solutions of the equations (5.1) to (5.5) to a sequence of linear equations. It is the so-called successive linearization method.

Suppose, for simplicity, the initial conditions to be fixed at the values of the stationary solution. Hence, keeping J_{io} fixed, $\underline{E}$, $\underline{H}$, $\underline{v}$, p, N will be functions of x and t, and also of η.

Assume that these functions can be expanded in power series of η. Rigorously speaking, the convergence of the series for any (x,t) should be proved, or, at least, we should estimate the error at the n-th approximation, to be defined in a moment.

Recalling the values assumed by $\underline{E}$, $\underline{H}$, N, p at $\eta = 0$, let:

$$\underline{E} = \underline{E}_o + \eta \underline{E}_1 + \eta^2 \underline{E}_2 + \ldots + \eta^n \underline{E}_n + \ldots , \tag{5.9}$$

$$\underline{H} = \eta \underline{H}_1 + \eta^2 \underline{H}_2 + \ldots + \eta^n \underline{H}_n + \ldots , \tag{5.10}$$

$$\underline{v} = \eta \underline{v}_1 + \eta^2 \underline{v}_2 + \ldots + \eta^n \underline{v}_n + \ldots , \tag{5.11}$$

$$N = N_o + \eta N_1 + \eta^2 N_2 + \ldots + \eta^n N_n + \ldots , \tag{5.12}$$

$$p = p_o + \eta p_1 + \eta^2 p_2 + \ldots + \eta^n p_n + \ldots , \tag{5.13}$$

where the terms multiplying powers of η depend only on x and t.

Inserting these into (5.1) to (5.7) and equating the terms independent of η , the stationary solution is recovered.

Equating the terms of first order in η, the following system is found:

$$\nabla \times \underline{H}_1 = \epsilon \frac{\partial \underline{E}_1}{\partial t} + N_o q \underline{v}_1 + \underline{J}_{io} , \quad \nabla \times \underline{E}_1 = = \mu \frac{\partial \underline{H}_1}{\partial t} ,$$

$$N_o m \frac{\partial \underline{v}_1}{\partial t} = N_o q \underline{E}_1 + N_1 q \underline{E}_o - \nabla p_1 , \tag{5.14}$$

$$\frac{\partial N_1}{\partial t} + \nabla . (N_o \underline{v}_1) = 0 , \quad p_1 = f'(N_o) N_1 .$$

These equations are called the first approximation.

If the plasma is a cold one, $p = 0$ and, hence, $\underline{E}_o = \underline{0}$.

Then the first three equations, respectively, coincide with (1.1) (in which $N = N_o(x)$), with (1.2) and with (2.1) (in which $\nu = H_o = 0$). The most widely used equations in the theory of electromagnetic fields in a plasma are, therefore, the first approximation of more general equations.

Let us, now, proceed to the second approximation equations. They are obtained by equating the terms in η^2. We have the system:

$$\nabla \times \underline{H}_2 = \epsilon \frac{\partial \underline{E}_2}{\partial t} + N_o q \underline{v}_2 + N_1 q \underline{v}_1 ,$$

$$\nabla \times \underline{E}_2 = - \mu \frac{\partial \underline{H}_2}{\partial t} ,$$

$$N_o m \left(\frac{\partial \underline{v}_2}{\partial t} + (\nabla \underline{v}_1) \underline{v}_1 \right) + N_1 m \frac{\partial \underline{v}_1}{\partial t} = N_o q \underline{E}_2 + N_1 q \underline{E}_1 \tag{5.15}$$

$$+ N_2 q \underline{E}_o + N_o q \mu \underline{v}_1 \times \underline{H}_1 - \nabla p_2 ,$$

$$\frac{\partial N_2}{\partial t} + \nabla.(N_o \underline{v}_2) + \nabla.(N_1 \underline{v}_1) = 0 ,$$

$$p_2 = f'(N_o) N_2 + \frac{f''(N_o)}{2} N_1^2 . \qquad (5.15)$$

Proceeding in this way, the n-th order approximation equations can be obtained. It is worth remarking that, in a cold plasma, we have $N_1 = 0$, where $\nabla.\underline{J}_{io}$ and ∇N_o are zero.

Taking the divergence of the third equation in (5.14), in which $p_1 = E_o = 0$, we find:

$$m \frac{\partial}{\partial t} (\nabla.(N_o \underline{v}_1)) = q \nabla.(N_o \underline{E}_1) = q N_o \nabla.\underline{E}_1 , \qquad (5.16)$$

and, taking the divergence of the first equation in (5.14),

$$\frac{\partial}{\partial t} \nabla.\underline{E}_1 = - \frac{1}{\epsilon} \nabla.(N_o q \underline{v}_1) - \nabla.\underline{J}_{io} . \qquad (5.17)$$

When x is fixed, (5.16) and (5.17) represent a system of two first order ordinary differential equations in $\nabla.(N_o \underline{v}_1)$ and $\nabla.\underline{E}_1$. Now, for $t = 0$, it has been assumed that $\nabla.(N_o \underline{v}_1) = 0$, $\underline{E}_1 = \underline{0}$. Then, at all points where $\nabla.\underline{J}_{io} = 0$ and for every instant of time, $\nabla.(N_o \underline{v}_1) = \nabla.\underline{E}_1 = 0$. Hence at these points, by the fourth equation in (5.14), we have $\frac{\partial N_1}{\partial t} = 0$. Therefore $N_1 = 0$, by the initial condition.

We remark that $N_1 = 0$ outside the sources. Inside the sources it could be $\nabla.\underline{J}_{io} \neq 0$. We must, however, remark also that, in general, there is no plasma, where the sources occur. Since $N_o q \underline{v}_1$ is, for the standard values of N_o, very small compared to J_{io}, its addition to J_{io} does not affect the electromagnetic field.

Then the first three equations of (5.15), if $\nabla N_o = \underline{0}$ everywhere, reduce to the equations considered in the preceding Section, except for the nonlinear terms, where $\underline{H}$ and $\underline{v}$ must be replaced by $\underline{H}_1$ and $\underline{v}_1$ (or, which is the same thing, by $\underline{v}_e$ and $\underline{H}_e$).

Therefore the arguments of the preceding Section can be repeated: in particular the last term in (4.17) reduces to $\frac{Nq}{2\epsilon} v_e^2$.

Hence we can assert the validity of the results of Sect. 4, for the second approximation for a cold plasma, and for equations more general than those employed in Sect. 4.

6. The linear equations of the first approximation lead to the introduction of some reciprocity theorems, in some cases of remarkable interest.

Since we will consider only theorems for quantities which are arbitrary functions of time (in particular, sinusoidal dependence), let us first recall the notion of a convolution product together with some of its properties. Then we shall look, setting aside for a moment consideration of the plasma, for a reciprocity theorem extending the classical Betti one to the dynamics of elastic bodies. Subsequently a reciprocity theorem will be described for the first approximation equations in a hot plasma.

As is known, given two continuous functions $A'(t)$ and $A''(t)$ of the variable t, their convolution product, denoted by $A'(t) * A''(t)$, is defined as follows:

$$A'(t) * A''(t) = \int_0^t A'(t-\tau)A''(\tau)\,d\tau . \tag{6.1}$$

The convolution product enjoys the same properties as products of numbers. Indeed we immediately verify that it is commutative and distributive. Furthermore it is associative, and, if it vanishes, at least one of the factors must vanish.

Let us, now, deduce some formulae, which will become useful in what follows. They hold under the assumptions that A' and A'' are of class C^1, with $A'(0) = A''(0) = 0$.

The extension to the general case yields no difficulties. Since the convolution product is commutative, we have:

$$\int_0^t A'(t-\tau)A''(\tau)\,d\tau = \int_0^t A'(\tau)A''(t-\tau)\,d\tau . \tag{6.2}$$

Differentiate (6.2) with respect to t. Then:

$$\int_0^t \frac{dA'(t-\tau)}{dt} A''(\tau)\,d\tau = \int_0^t A'(\tau)\frac{dA''(t-\tau)}{dt}\,d\tau .$$

That is, when $A'(0) = A''(0) = 0$, we get:

$$\frac{d}{dt} A'(t) * A''(t) = A'(t) * \frac{d}{dt} A''(t) . \qquad (6.3)$$

If, in addition, we have $\frac{dA'(0)}{dt} = \frac{dA''(0)}{dt} = 0$, A' and A'' being of class C^2, then the following formula holds:

$$\frac{d^2 A'(t)}{dt^2} * A''(t) = A'(t) * \frac{d^2 A''(t)}{dt^2} . \qquad (6.4)$$

Given two vectors $\underline{a}'(t)$, $\underline{a}''(t)$, with components $a'_i(t)$, $a''_i(t)$, $i = 1, 2, 3$, which are continuous functions of t, the convolution product between a'(t) and a''(t) is defined by

$$\underline{a}'(t) * \underline{a}''(t) = \int_0^t \underline{a}'(t-\tau).\underline{a}''(\tau)\, d\tau = \int_0^t \underline{a}'_i(t-\tau).\underline{a}''_i(\tau)\, d\tau . \qquad (6.5)$$

It is easy to see that the convolution product between vectors is commutative and distributive, and that the properties expressed by (6.3) and (6.4) are still valid.

5.7 Consider an elastic body occupying a domain $\mathscr{D}$ with boundary σ. In the framework of the linear theory of elasticity, let us consider two motions: the first one generated by the mass force $\rho\underline{F}'$ (ρ the density) and by the surface force $\underline{R}'$, and the second one by the forces $\rho\underline{F}''$ and $\underline{R}''$.

Let $\underline{u}'$ be the displacement vector of the first motion and $\underline{u}''$ that of the second one. The mass forces and the displacements depend in general on $x \in \mathscr{D}$ and t; the surface forces depend on $x \in \sigma$ and t. Let us take as zero both the displacements and their time derivatives at the initial time. This is a simplifying hypothesis: however it is not difficult to extend the theorem to the general case of nonzero initial conditions.

Now let u'_i, u''_i be the cartesian components of the displacements $\underline{u}'$, $\underline{u}''$, and σ'_{ij}, σ''_{ij} be the components of the stress tensors of the two motions.

The fundamental equation of an elastic body, when written for the first motion, gives:

$$\rho \frac{\partial^2 u'_i}{\partial t^2} = \rho F'_i + \sigma'_{ij/j} ; \qquad (7.1)$$

(as usual, summation over the repeated index j is implicitly assumed).

By replacing the upper case index ' by '', we get the equation for the second motion.

Now, if ϵ'_{hk}, ϵ''_{hk} denote the components of the strain tensor of the two motions, by Hooke's law we have:

$$\sigma'_{ij} = c_{ijhk}\,\epsilon'_{hk} \tag{7.2}$$

(the indices h and k are summed); a similar formula is applicable for the second motion.

The c_{ijhk} are the elastic constants of the body. They have the following properties:

$$c_{ijhk} = c_{hkij}\,. \tag{7.3}$$

Let us, now, consider the expression in which the indices i and j are summed, i.e. the sum of the convolution products between σ'_{ij} and the corresponding ϵ''_{ij}. We have:

$$\sigma'_{ij} * \epsilon''_{ij} = c_{ijhk}\,\epsilon'_{hk} * \epsilon''_{ij}\,. \tag{7.4}$$

In the right-hand side there is a summation over the indices i, j, h, k. Interchanging the indices i, j with h, k, we have, on account of (7.3),

$$\sigma'_{ij} * \epsilon''_{ij} = c_{hkij}\,\epsilon'_{ij} * \epsilon''_{hk} = c_{ijhk}\,\epsilon''_{hk} * \epsilon'_{ij}\,. \tag{7.5}$$

Now, interchanging in (7.4) the upper case indices, we obtain, by (7.3), the right-hand side of (7.5). Hence we have:

$$\sigma'_{ij} * \epsilon''_{ij} = \sigma''_{ij} * \epsilon'_{ij}\,. \tag{7.6}$$

Furthermore, by (6.4),

$$\frac{\partial^2 u'_i}{\partial t^2} * u''_i = \frac{\partial^2 u''_i}{\partial t^2} * u'_i\,. \tag{7.7}$$

Now perform the convolution product between (7.1) and u''_i. A sum over i being implicitly assumed, we get:

$$\frac{\partial^2 u'_i}{\partial t^2} * u''_i = \rho F'_i * u''_i + \sigma'_{ij/j} * u''_i = \rho F'_i * u''_i + \tag{7.8}$$
$$+ (\sigma'_{ij} * u''_i)_{/j} - \sigma'_{ij} * \epsilon''_{ij} \ .$$

Integrating over the volume S of $\mathscr{D}$, applying Green's theorem and recalling that, if n_i are the components of the outward normal $\underline{n}$ to σ, we have:

$$R'_i = \sigma'_{ij} n_j \ , \tag{7.9}$$

we get:

$$\int_S \rho \frac{\partial^2 u'_i}{\partial t^2} * u'' dS + \int_S \sigma'_{ij} * \epsilon''_{ij} dS = \int_S \rho F'_i * u''_i dS + \int_\sigma \sigma'_{ij} n_j * u''_i d\sigma =$$
$$= \int_S \rho F'_i * u''_i dS + \int_\sigma R'_i * u''_i d\sigma \ . \tag{7.10}$$

Now, by (7.6) and (7.7), the left-hand side is left invariant by an interchange of the upper case indices. Hence the right-hand side of (7.9) will not change, and, by (6.5), we get:

$$\int_S \rho \underline{F}' * \underline{u}'' dS + \int_\sigma \underline{R}' * \underline{u}'' d\sigma = \int_S \rho \underline{F}'' * \underline{u}' dS + \int_\sigma \underline{R}'' * \underline{u}' d\sigma \ , \tag{7.11}$$

which represents the reciprocity relation of elastodynamics we were looking for, completely analogous to the Betti theorem, except for the fact that the ordinary product is replaced by a convolution product.

Consider, now, the particular case of a synchronous force. That is, let

$$\rho \underline{F}'(x,t) = G(t) \underline{a}'(x) \ , \qquad \underline{R}' = G(t) \underline{b}'(x) \ , \tag{7.12}$$

$$\underline{F}''(x,t) = G(t) \underline{a}''(x) \ , \qquad \underline{R}'' = G(t) \underline{b}''(x) \ , \tag{7.13}$$

i.e. the forces can be decomposed into an overall function of time $G(t)$, and to different functions of the space variables. Substituting in (7.11) we get:

$$G(t) * \left[\int_S \underline{a}'.\underline{u}''dS + \int_\sigma \underline{b}'.\underline{u}''d\sigma - \int_S \underline{a}''.\underline{u}'dS - \int_\sigma \underline{b}''.\underline{u}'d\sigma \right] = 0. \quad (7.14)$$

Now, by the properties of the convolution product, since $G(t) \neq 0$, the term within square brackets vanishes. Multiplying this term by $G(t)$, by (7.12) and (7.13), we have at every instant:

$$\int_S \underline{F}'.\underline{u}''dS + \int_\sigma \underline{R}'.\underline{u}''d\sigma = \int_S \underline{F}''.\underline{u}'dS + \int_\sigma \underline{R}''.\underline{u}'d\sigma, \quad (7.15)$$

That is, when the forces are synchronous and the initial conditions are zero, the Betti theorem in its usual form holds even in the case of elastodynamics [12].

5.8 Let us consider again the plasma, and let us deduce a reciprocity theorem for the first order equations (5.14), where, to simplify the notations, the indices will be dropped, except for N_o, $\underline{E}_o$ and N_1. In place of $\underline{J}_{io}$ let us write again $\underline{J}_i$.

$$\nabla \times \underline{H} = \epsilon \frac{\partial \underline{E}}{\partial t} + N_o q \underline{v} + \underline{J}_i, \quad (8.1)$$

$$\nabla \times \underline{E} = -\mu \frac{\partial \underline{H}}{\partial t}, \quad (8.2)$$

$$N_o m \frac{\partial \underline{v}}{\partial t} = N_o q \underline{E} - \nabla p + N_1 q \underline{E}_o, \quad (8.3)$$

$$\frac{\partial N_1}{\partial t} + \nabla.(N_o \underline{v}) = 0, \quad (8.4)$$

$$p = f'(N_o) N_1. \quad (8.5)$$

The range of validity of these equations is assumed to be the whole space $\mathscr{D}_\infty$; however, let the region $\mathscr{D}_i$ containing the sources be bounded. Thus in $\mathscr{D}_\infty \cap \mathscr{D}_i$, J_i vanishes. N_o is assumed to depend on x, and vanishes on $\mathscr{D}_i$ and on its boundary σ. Consequently p also vanishes on $\mathscr{D}_i$ and on σ. Furthermore, we shall assume zero initial conditions.

By a uniqueness theorem [11] generalizing that of Sect. 5.3, it can be proved that, outside a sphere of suitably large radius, $\underline{E}$, $\underline{H}$, and $\underline{v}$ vanish.

Consider, now, two fields $\underline{E}'$, $\underline{H}'$, $\underline{v}'$, N', p'; $\underline{E}''$, $\underline{H}''$, $\underline{v}''$, N'', p'', generated by the currents $\underline{J}'_i$, $\underline{J}''_i$, respectively. Let these fields vanish at the initial time. Of course both fields are solutions of (8.1) to (8.5).

Consider, now, a point x outside $\mathscr{D}_i$ and, taking into account (8.1), (8.2), we have the following expression:

$$\nabla . \int_0^t \underline{E}'(t-\tau) \times \underline{H}''(\tau)\, d\tau = \int_0^t (\nabla \times \underline{E}'(t-\tau) . \underline{H}''(\tau)$$

$$- \underline{E}'(t-\tau) . \nabla \times \underline{H}''(\tau))\, d\tau = \tag{8.6}$$

$$= -\mu \frac{\partial \underline{H}'}{\partial t} * \underline{H}'' - \epsilon\, \underline{E}' * \frac{\partial \underline{E}''}{\partial t} - N_o q\, \underline{E}' * \underline{v}'' .$$

Take the convolution product of $\underline{v}''$ and (8.3) (with $\underline{v}$, $\underline{E}$, p, N replaced by $\underline{v}'$, $\underline{E}'$, p', N').

We have:

$$N_o m \frac{\partial \underline{v}'}{\partial t} * \underline{v}'' = N_o q\, \underline{E}' * \underline{v}'' - \nabla p' * \underline{v}'' + N'_1 q\, E'_o * \underline{v}'' , \tag{8.7}$$

which can be rewritten as

$$\nabla . (p' * \underline{v}'') = - N_o m \frac{\partial \underline{v}'}{\partial t} * \underline{v}'' + N_o q\, \underline{E}' * \underline{v}'' + p' * \nabla . \underline{v}''$$

$$+ N'_1 q\, \underline{E}'_o * \underline{v}'' . \tag{8.8}$$

Adding (8.8) to (8.6), we get:

$$\nabla . \left\{ \int_0^t (\underline{E}'(t-\tau) \times \underline{H}''(\tau))\, d\tau + p' * \underline{v}'' \right\} = -\mu \frac{\partial \underline{H}'}{\partial t} * \underline{H}''$$

$$- \epsilon\, \underline{E}' * \frac{\partial \underline{E}''}{\partial t} + N'_1 q\, \underline{E}'_o * \underline{v}'' - N_o m \frac{\partial \underline{v}'}{\partial t} * \underline{v}'' + p' * \nabla . \underline{v}'' . \tag{8.9}$$

Now, taking into account (5.7) and the second equation of (5.8), we have:

$$p' * \nabla . \underline{v}'' + N'_1 q\, \underline{E}'_o * \underline{v}'' = p' * \nabla . \underline{v}'' + \frac{f'(N_o)}{N_o} \nabla N_o N'_1 * \underline{v}'' . \tag{8.10}$$

Now, replacing N'_1 and $\underline{v}'$ by N''_1 and $\underline{v}''$, (8.4) yields

$$\nabla.\underline{v}'' = -\frac{1}{N_o}\frac{\partial N''_1}{\partial t} - \frac{\nabla N_o}{N_o}.\underline{v}''. \tag{8.11}$$

Inserting this into (8.10), and recalling that, by (8.5), $p' = f'(N_o)N'_1$ (we remark that $f'(N_o)$ is the derivative of $f(N)$ with respect to N, at $N = N_o$; $f'(N_o)$ does not depend on t and is the same for both fields), we obtain:

$$\begin{aligned} p' * \nabla.\underline{v}'' + N'_1\, q\, \underline{E}'_o * \underline{v}'' &= -\frac{f'(N_o)}{N_o} N'_1 * \frac{\partial N''_1}{\partial t} \\ -\frac{f'(N_o)}{N_o}\nabla N_o . \underline{v}'' * N'_1 + \frac{f'(N_o)}{N_o}\nabla N_o . N'_1 * \underline{v}'' &= -\frac{f'(N_o)}{N_o} N'_1 * \frac{\partial N''_1}{\partial t}. \end{aligned} \tag{8.12}$$

Equation (8.12) shows, when inserted into (8.9) and on account of (6.3), that the right-hand side of (8.9) is left invariant by an interchange between $\underline{v}'$, . . . and $\underline{v}''$, ... We thus obtain a relation which, essentially, represents the reciprocity theorem:

$$\begin{aligned} &\nabla.\left\{\int_0^t \underline{E}'(t-\tau)\times\underline{H}''(\tau)\,d\tau + p' * v''\right\} = \\ &= \nabla.\left\{\int_0^t \underline{E}''(t-\tau)\times\underline{H}'(\tau)\,d\tau + p'' * \underline{v}'\right\}. \end{aligned} \tag{8.13}$$

Consider the region $\mathscr{D}$ containing the sources. By our assumptions on $\mathscr{D}$, N_o vanishes and J_i does not. Repeating the above arguments, we find:

$$\begin{aligned} &\nabla.\int_0^t \underline{E}'(t-\tau)\times\underline{H}''(\tau)\,d\tau - \underline{J}''_i * \underline{E}' = \\ &= \nabla.\int_0^t \underline{E}''(t-\tau)\times H'(\tau)\,d\tau - \underline{J}'_i * \underline{E}''. \end{aligned} \tag{8.14}$$

Now integrate (8.14) over the volume S of $\mathscr{D}$, bounded by σ, whose outward normal is $\underline{n}$, and (8.13) on the domain $\mathscr{D}_1$, bounded by the surfaces Σ and σ. Summing these equations, by Green's theorem we have two integrals over σ that cancel each other, since on σ $p' = p'' = 0$ and the tangential components of $\underline{E}'$, $\underline{E}''$, $\underline{H}'$, $\underline{H}''$ are equal, so that the components along $\underline{n}$ have opposite signs. Thus there remains, if $\underline{n}$ is the outward normal to Σ,

$$\int_S - \underline{J}'_i * \underline{E}''\,dS + \int_\Sigma \left[\int_0^t \underline{E}''(t-\tau) \times \underline{H}'(\tau)\,d\tau \right] . \underline{n}\,d\Sigma + \int_\Sigma p'' * \underline{v}' . \underline{n}\,d\Sigma$$

$$= \int_S - \underline{J}''_i * \underline{E}'\,dS + \int_\Sigma \left[\int_0^t \underline{E}'(t-\tau) \times \underline{H}''(\tau)\,d\tau \right] . \underline{n}\,d\Sigma + \int_\Sigma p' * \underline{v}'' . \underline{n}\,d\Sigma . \qquad (8.15)$$

If Σ is chosen at a suitably large distance from the sources, so that the fields vanish on it, the integrals over Σ are zero and there remains

$$\int_S \underline{J}'_i * \underline{E}''\,dS = \int_S \underline{J}''_i * \underline{E}'\,dS , \qquad (8.16)$$

which yields the reciprocity theorem in the form most convenient for applications.

6 Propagation of electromagnetic waves in a nonlinear dielectric

6.1 In Chapter 1 nonlinear dielectric media have been mentioned. If an electromagnetic (in particular a light) wave propagates in these media, some interesting phenomena take place, which have given rise to a new chapter of physics, namely nonlinear optics.

The dielectric media encountered in nonlinear optics are, in general, non-isotropic. The relation between the displacement vector $\underline{D}$ and the dielectric field $\underline{E}$ is expressed by (3.5) of Chapter 2:

$$\underline{D} = \underline{D}(\underline{E}) , \tag{1.1}$$

whereas we can still assume the validity of (3.2) and (3.3) of Chapter 2 (with $\underline{E}_i = \underline{0}$ because the sources of the electromagnetic field lie outside the nonlinear dielectrics). That is, we have:

$$\underline{B} = \mu \underline{H} ; \tag{1.2}$$

$$\underline{J} = \gamma \underline{E} , \tag{1.3}$$

where μ in practice coincides with the vacuum permeability. γ is taken to be a scalar quantity.

To investigate the properties of (1.1), it is convenient to introduce an orthogonal coordinate system, y_i, i = 1, 2, 3; $\underline{e}_i$ is the unit vector along the axis y_i. Let D_i, D_h, E_i, E_h, i, h = 1, 2, 3, be the components of $\underline{D}$ and $\underline{E}$ along the axes. Then (1.1) becomes

$$D_i = D_i (E_j) . \tag{1.4}$$

Analogous assumptions and arguments (it is enough to interchange the magnetic energy with the electric one, $\underline{B}$ with $\underline{D}$, $\underline{H}$ with $\underline{E}$) lead to relations corresponding to (4.6) of Chapter 2 (note that j is replaced by i):

$$\frac{\partial D_i}{\partial E_k} = \frac{\partial D_k}{\partial E_i} . \tag{1.5}$$

A usual assumption of nonlinear Optics is the following one: D_i is a sum of a linear function of the components of the electric field and of a quadratic one, i.e. we write:

$$D_i = \epsilon_{ik} E_k + d_{ijk} E_j E_k , \tag{1.6}$$

where in the right-hand side summation is over the indices k and j.

It is easily proved that we can always take $d_{ijk} = d_{ikj}$. Furthermore, by (1.5), we have:

$$\epsilon_{ik} + d_{ijk} E_j = \epsilon_{ki} + d_{kji} E_j \tag{1.7}$$

and, since E_j is arbitrary, we obtain:

$$\epsilon_{ik} = \epsilon_{ki} , \quad d_{ijk} = d_{kji} . \tag{1.8}$$

i.e. the tensors ϵ_{ik} and d_{ikj} are symmetric, the second one with respect to all three indices. (It is worth remarking that these results assume $\underline{D}(t)$ to be a function only of $\underline{E}$, computed at the same instant t: to account for dispersive phenomena, as will be mentioned later, this hypothesis has to be modified, so that (1.8) are not always in agreement with experiments).

Let us remark that it is always possible to choose the axes in such a way that $\epsilon_{ik} = 0$ for $i \neq k$. Finally we remark that, if the dielectric is nonlinear, but isotropic, the same arguments used to deduce (4.8) of Chapter 2 lead to the formula:

$$\underline{D} = \epsilon(E^2)\underline{E} . \tag{1.9}$$

6.2 Let us, now, consider a plane parallel plate of thickness s, formed by a dielectric of the type just described. For convenience, replace the coordinate system $Oy_1 y_2 y_3$ by a system Oxyz, such that one face of the plate lies on the xy plane, and the z axis is oriented in such a way that the opposite face of the plate lies on the plane with the equation $z = s$.

Outside the plate (i.e. for $z < 0$ or $z > s$) let the medium coincide with the vacuum, or with air.

In the halfspace $z < 0$ let there be located a source generating a plane electromagnetic wave propagating along the z axis in the positive direction.

Beyond the generated wave, called incident wave, in the halfspace $z < 0$ there will be a wave reflected by the plane $z = 0$. In the halfspace $z > s$ there will be only a transmitted wave, and within the plate two waves, propagating along the z axis in opposite directions.

Assume that the source emits polarized light, and that the plate is formed in such a way that the axes x and y can be chosen so as to express the fields $\underline{E}$ and $\underline{H}$ through the formulae:

$$\underline{E} = E\underline{i}\,, \quad \underline{H} = H\underline{j}\,, \tag{2.1}$$

where E and H are scalar quantities depending on z and t. That is, let us assume the wave to be linearly polarized in the whole space.

By (1.2), (1.3) and (2.1), $\underline{B}$ is parallel to $\underline{H}$, the current density $\underline{J}$ is parallel to $\underline{E}$ and $\underline{i}$ is parallel to $\underline{E}$. Furthermore, let us assume $\underline{D}$ to be parallel to $\underline{E}$ within the plate, so that

$$\underline{D} = D(E)\underline{i}\,. \tag{2.2}$$

The Maxwell equations within the plate are in this case (see (2.1), (2.2) of Chapter 2 and (2.2), (1.2), (1.3)):

$$-\frac{\partial H}{\partial z} = \frac{\partial D}{\partial t} + \gamma E = \frac{\partial D}{\partial E}\frac{\partial E}{\partial t} + \gamma E \tag{2.3}$$

$$-\frac{\partial E}{\partial z} = \mu\frac{\partial H}{\partial t}\,. \tag{2.4}$$

Differentiating (2.3) with respect to t, and (2.4) with respect to z, multiplying (2.3) by μ and then subtracting, we find:

$$\frac{\partial^2 E}{\partial z^2} = \mu\frac{\partial^2 D}{\partial t^2} + \gamma\mu\frac{\partial E}{\partial t}\,. \tag{2.5}$$

6.3 Equations (2.3) and (2.4) then hold within the plate, where it is assumed there is a known function D(E) relating D to E.

By solving these equations, we find both the reflected and the transmitted wave. We have, however, to establish conditions for $z = 0$ and $z = s$, which make the solutions of (2.3) and (2.4) unique.

To this end, let us remark that, for $z < 0$, we have, as already seen, both the incident and the reflected wave.

Let $E_i(z,t)\underline{i}$, $H_i(z,t)\underline{j}$ be the electric and the magnetic fields, respectively, of the incident wave, and $E_r(z,t)\underline{i}$, $H_r(z,t)\underline{j}$ be those of the reflected wave. Now let $E(z,t)\underline{i}$, $H(z,t)\underline{j}$ be the fields within the plate.

Since the tangential components of the electric and of the magnetic field must be continuous at the $z = 0$ face (Chapter 2, Sect. 2), these components being the fields themselves, and since in the halfspace $z < 0$ the total field is the sum of the fields of the incident and reflected wave, we have:

$$E_i(0,t) + E_r(0,t) = E(0,t) . \tag{3.1}$$

$$H_i(0,t) + H_r(0,t) = H(0,t) . \tag{3.2}$$

Now, if ϵ_o is the vacuum dielectric constant and μ is the same at all points of space, we have (see [13], Chapter 2, equations (17), (18), (19)):

$$H_i(z,t) = \sqrt{\frac{\epsilon_o}{\mu}}\, E_i(z,t) ; \tag{3.3}$$

$$H_r(z,t) = -\sqrt{\frac{\epsilon_o}{\mu}}\, E_r(z,t) . \tag{3.4}$$

Inserting (3.3) and (3.4) in (3.2), and adding the resulting equations after multiplication of (3.1) by $\sqrt{\frac{\epsilon_o}{\mu}}$, we get:

$$\sqrt{\frac{\epsilon_o}{\mu}}\, E(0,t) + H(0,t) = \sqrt{\frac{\epsilon_o}{\mu}}\, E_i(0,t) + H_i(0,t) . \tag{3.5}$$

In the halfspace $z > s$ there is only a transmitted wave, whose fields are $E_t(z,t)\underline{i}$ and $H_t(z,t)\underline{j} = \sqrt{\frac{\epsilon_o}{\mu}}\, E_t(z,t)\underline{j}$.

By the above argument, the following equations are obtained:

$$E(z,t) = E_t(z,t), \qquad H(z,t) = \sqrt{\frac{\epsilon_o}{\mu}}\, E_t(z,t),$$

whence, eliminating E_t, we find that

$$E(z,t) - \sqrt{\frac{\mu}{\epsilon_o}}\, H(z,t) = 0. \tag{3.6}$$

Equations (3.5) and (3.6) (E_i and H_i are assumed to be known) are the boundary conditions to be associated with (2.3) and (2.4).

The problem just described is contained within the framework of a general theory of partial differential equations developed by Professor Cesari. He proved that, when s is not too large, there holds a theorem that asserts the existence, uniqueness and continuous dependence on the data for (2.3), (2.4), with the boundary conditions (3.5), (3.6). ([14], [15], [16]).

Professor Bassanini [18] showed later that the thickness s for which Cesari's results hold is larger than the thickness used in experiments.

6.4 Let us, now, describe a solution, in some sense implicit, of (2.5) with $\gamma = 0$. This solution, as we shall see later on, is of some usefulness.

To find this solution, suppose $\partial D/\partial E$ to be positive, which is the same thing, as could be proved, as assuming the tensor $\partial D_i/\partial E_k$ to be positive. By analogy with the arguments of Chapter 2, Sect. 2.4, $\partial D/\partial E$ corresponds to ϵ of the linear case, and hence a natural situation is:

$$p^2(E) = \frac{\partial D}{\partial E}\,\mu, \tag{4.1}$$

and it is natural to look for a solution of the type:

$$E = G(t - p(E)(z - z_o) + \alpha), \tag{4.2}$$

where α is a constant, z_o a point inside the plate and $G(u)$ an arbitrary function of $u = t - p(E)(z - z_o) + \alpha$.

Now, denoting by $G'(u)$ the derivative of $G(u)$ with respect to u, we have:

$$\frac{\partial E}{\partial z} = G'(t - p(E)(z - z_o) + \alpha)\;[-p(E) - (z - z_o)\frac{\partial p}{\partial E}\;\frac{\partial E}{\partial z}], \tag{4.3}$$

$$\frac{\partial E}{\partial t} = G'(t - p(E)(z - z_o) + \alpha)\,[1 - (z - z_o)\,\frac{\partial p}{\partial E}\,\frac{\partial E}{\partial t}]\,, \tag{4.4}$$

$$\frac{\partial E}{\partial z}(1 + G'(u)(z - z_o)\,\frac{\partial p}{\partial E}) = -\,G'(u)p(E)\,, \tag{4.5}$$

$$\frac{\partial E}{\partial t}(1 + G'(u)\,(z - z_o)\,\frac{\partial p}{\partial E}) = G'(u)\,. \tag{4.5'}$$

Now, if the factor multiplying $\partial E/\partial z$ in (4.5) and $\partial E/\partial t$ in (4.5') (denoted for short by m) does not vanish, we immediately have:

$$\frac{\partial E}{\partial z} = -\,p(E)\,\frac{\partial E}{\partial t}\,. \tag{4.6}$$

We remark that m certainly does not vanish for $u \in (-h, h)$ and z belonging to some neighbourhood of z_o, which is large because $\partial p/\partial E$ is actually very small. The neighbourhood contains all values of the variables such that $\partial E/\partial t$ and $\partial E/\partial z$ remain bounded, so that if these derivatives are assumed to be always and everywhere bounded we have $m \neq 0$ in any case and (4.6) holds.

Taking into account (4.6) we have:

$$\frac{\partial^2 E}{\partial z^2} = -\,\frac{\partial}{\partial z}\,(p(E)\,\frac{\partial E}{\partial t}) = -\,p(E)\,\frac{\partial^2 E}{\partial z\,\partial t} - \frac{\partial p(E)}{\partial E}\,\frac{\partial E}{\partial z}\,\frac{\partial E}{\partial t}\,, \tag{4.7}$$

$$\begin{aligned}\mu\,\frac{\partial^2 D}{\partial t^2} &= \frac{\partial}{\partial t}\,(\mu\,\frac{\partial D}{\partial E}\,\frac{\partial E}{\partial t}) = \frac{\partial}{\partial t}\,(p^2(E)\,\frac{\partial E}{\partial t}) = -\,\frac{\partial}{\partial t}\,(p(E)\,\frac{\partial E}{\partial z}) = \\ &= -\,p(E)\,\frac{\partial^2 E}{\partial z\,\partial t} - \frac{\partial p(E)}{\partial E}\,\frac{\partial E}{\partial t}\,\frac{\partial E}{\partial z}\,.\end{aligned} \tag{4.8}$$

Comparing (4.7) with (4.8), it is easily seen that (4.2) satisfies (2.5).

On account of (4.1) formulae (2.3) and (2.4) yield:

$$\frac{\partial H}{\partial z} = -\,\frac{p^2}{\mu}\,\frac{\partial E}{\partial t}\,, \qquad \frac{\partial H}{\partial t} = -\,\frac{1}{\mu}\,\frac{\partial E}{\partial z}\,. \tag{4.9}$$

From these equations, which are compatible by (2.5), we can find H. Let us put:

$$H = \frac{1}{\mu} \int p(E)\, dE \,. \tag{4.10}$$

Differentiating and taking into account (4.6) we have:

$$\frac{\partial H}{\partial z} = \frac{p(E)}{\mu} \frac{\partial E}{\partial z} = - \frac{p^2}{\mu} \frac{\partial E}{\partial t} \,,$$

$$\frac{\partial H}{\partial t} = \frac{p(E)}{\mu} \frac{\partial E}{\partial t} = - \frac{1}{\mu} \frac{\partial E}{\partial z} \,,$$

so that (4.10) satisfies (4.9).

6.5 Let us now try to solve (2.3) and (2.4), or better (2.5), by means of an approximate method often applied in nonlinear Optics. We shall try to clarify it also by applying the approximate method of nonlinear mechanics known as the Kryloff-Bogoliubov-Mitropolsky method [17] (hereafter, the KBM method).

It is worth stating from the start that, to the best of my knowledge, the validity of this method has been proved for ordinary, but not for partial, differential equations. It is, however, reasonable to assume its validity also for the partial differential equations. This assumption will be confirmed, at least partially, in Sect. 9, and later on we shall comment further on its validity. In any case, it remains a heuristic method which so far has led to results in agreement with the physical intuition.

To simplify the computations, let us assume the thickness s of the nonlinear plate equal to infinity; that is that the nonlinear medium fills the halfspace $z > 0$. In this case the arguments on the plane $z = 0$ remain valid, i.e. equations (3.1), (3.2), (3.3), (3.4) (some of them can be replaced by (3.5)). However (3.6) becomes meaningless, and it has to be replaced by the hypothesis (the medium is supposed to be homogeneous for $z > 0$ and the source is in the halfspace $z < 0$ so that there are no reflected waves for $z > 0$) that for $z > 0$ there are only waves propagating along the positive direction of the z axis.

Since the KBM method holds for weak nonlinearities, set:

$$D(E) = \epsilon E + \eta F(E) \,, \tag{5.1}$$

where ϵ is some constant, $F(E)$ some nonlinear function of E, and η some very small number. The term $\eta F(E)$ can be regarded as a small perturbation of the usual relation between D and E, so that ϵ is the dielectric constant of the medium if the nonlinearity can be neglected.

We shall also assume the conductivity γ to be very small, and to be of the order of η, for $z > 0$. Therefore we can write $\gamma = \eta\gamma_1$, where γ_1 is a finite quantity. Then (2.5) becomes:

$$\frac{\partial^2 E}{\partial z^2} = \epsilon\mu \frac{\partial^2 E}{\partial t^2} + \mu\eta \frac{\partial^2 F(E)}{\partial t^2} + \eta\mu\gamma_1 \frac{\partial E}{\partial t} . \tag{5.2}$$

Let us suppose, as happens in applications, that the field source is a sinusoidal function of frequency ω. Now, if $\eta = 0$, (5.2) becomes linear, so that there would be a sinusoidal wave propagating along the positive direction of the z axis. Hence the value of the electric field would be given by:

$$E = Ae^{i\theta} + A^* e^{-i\theta} , \tag{5.3}$$

$$\theta = \omega(t - \sqrt{\epsilon\mu}\, z), \tag{5.3'}$$

where A is some complex number, and A^* is its complex conjugate.

The occurrence of the nonlinear term leads to the natural assumption that E is a periodic function of t with period $2\pi/\omega$, so that it can be expanded in a Fourier series or, more generally, in a series of waves propagating along the z axis, in the positive direction. On the other hand, also by the occurrence of the conductivity, the amplitude of such waves is presumably a non-constant function of z. Hence, in the expression for E, the factor A will not be assumed constant, but to vary with z, so that we will write:

$$A = A(z) = a(z)\, e^{i\varphi(z)} , \tag{5.4}$$

where $a(z)$[1] is the wave amplitude at the point z, and $\varphi(z)$ is its phase. Sometimes, for brevity, we shall simply write A, a, φ. Furthermore, as already mentioned, there will be other waves of frequency $n\omega$ (coming from the Fourier expansion), but of phase that in general will be a multiple of φ. In addition these waves arise by the occurrence of the terms in η, so that they will naturally be of the order of η. Hence we shall write:

---

(1) Strictly speaking, a is a one-half the wave amplitude.

$$E = (A(z)e^{i\theta} + A^*(z)e^{-i\theta}) + \eta\, \Sigma'_n\, (c_n e^{in(\theta+\varphi)} + c^*_n e^{-in(\theta+\varphi)}) \tag{5.5}$$

where, as already mentioned, $A^*(z)$ is the complex conjugate of $A(z)$, c_n is a function of a and z, and c^*_n is its complex conjugate.[(2)] Σ' denotes the sum over all n ranging from 0 to ∞, except 1, already extracted in (5.5). Moreover, this term of frequency ω is called the principal term or the principal wave of E, and the remaining ones are called higher order harmonics, or n-th order harmonics, since they have frequency $n\omega$. It is convenient to rewrite (5.5) in the following way:

$$E = A(z)e^{i\theta} + \eta\, \Sigma'_n\, c_n\, e^{in(\theta+\varphi)} + \text{c.c.} \tag{5.6}$$

where c.c denotes the complex conjugate of any term appearing in the right-hand side. It is worth remarking that the derivatives with respect to the real variables z and t of the c.c terms are none other than the complex conjugate of the derivative of the terms explicitly written in the right-hand side of (5.6).

Let us now come back to the assumption that A varies with z or, equivalently, that its modulus a and its argument φ vary with z. It is a natural assumption that this dependence is an increasing function of η, because if $\eta \to 0$ the medium becomes linear and a and φ become constant.

Furthermore the greater is the amplitude a of the principal wave the greater are the variations of a and φ. Hence, in agreement with the KBM method, it is natural to put:

$$\frac{da}{dz} = \eta\, B(a)\,, \qquad \frac{d\varphi}{dz} = \eta\, C(a) \tag{5.7}$$

where $B(a)$ and $C(a)$ are C^1 functions of a, bounded for bounded a. Then we have:

$$\frac{da}{dz} = O(\eta)\,, \qquad \frac{d\varphi}{dz} = O(\eta) \tag{5.8}$$

Hence, by (5.4):

---

(2) It is worth remarking that the present arguments and the further ones of the same type are intended only as a justification (and not a proof) of (5.5) and its consequences.

$$\frac{dA}{dz} = e^{i\varphi}\left(\frac{da}{dz} + ia\frac{d\varphi}{dz}\right) = \eta e^{i\varphi}(B(a) + ia\,C(a)) \tag{5.9}$$

and differentiating again with respect to z, taking into account (5.7), we easily get:

$$\frac{d^2A}{dz^2} = \eta e^{i\varphi}\left(\frac{dB}{da}\frac{da}{dz} + i\frac{da}{dz}C(a) + ia\frac{dC}{da}\frac{da}{dz}\right) +$$

$$+ \eta e^{i\varphi}\frac{d\varphi}{dz}(B(a) + ia\,C(a)) = O(\eta^2). \tag{5.10}$$

We notice now that $c_n = c_n(z,a)$. Then, recalling (5.6), we have:

$$\eta\frac{d}{dz}(c_n e^{in\varphi}) = \eta\left[\left(\frac{\partial c_n}{\partial z} + \frac{\partial c_n}{\partial a}\frac{da}{dz}\right)e^{in\varphi} + c_n e^{in\varphi}\, in\frac{d\varphi}{dz}\right] =$$

$$= \eta\frac{\partial c_n}{\partial z}e^{in\varphi} + O(\eta^2). \tag{5.11}$$

That is, up to terms of order $O(\eta^2)$, $\eta c_n e^{in\varphi}$, and in particular ηc_n, can be differentiated considering a and φ as constants. Let us now proceed to compute the various terms of (5.2). All terms of order η^2 are collected in the symbol $O(\eta^2)$. We could verify that the derivatives with respect to both z and t of terms $O(\eta^2)$ are always $O(\eta^2)$. For the sake of brevity this verification will be omitted. Then, by twice differentiating (5.6) with respect to z and t, taking into account (5.10) and (5.11), and noticing that, by (5.3'),

$$\frac{\partial\theta}{\partial z} = -\omega\sqrt{\epsilon\mu}, \quad \frac{\partial\theta}{\partial t} = \omega.$$

We have:

$$\frac{\partial E}{\partial z} = -i\omega\sqrt{\epsilon\mu}\,A e^{i\theta} + \eta\,\Sigma'_n\left(-in\omega\sqrt{\epsilon\mu}\,c_n + \frac{dc_n}{dz}\right)e^{in(\theta+\varphi)} + \frac{dA}{dz}e^{i\theta} +$$

$$+ c.c. + O(\eta^2) \tag{5.12}$$

$$\frac{\partial^2 E}{\partial z^2} = -\omega^2\epsilon\mu A e^{i\theta} - 2i\omega\sqrt{\epsilon\mu}\frac{dA}{dz}e^{i\theta} + \eta\,\Sigma'_n\Big(-n^2\omega^2\epsilon\mu c_n -$$

$$- 2in\omega\sqrt{\epsilon\mu}\frac{dc_n}{dz} + \frac{d^2c_n}{dz^2}\Big)e^{in(\theta+\varphi)} + O(\eta^2) + c.c. \tag{5.13}$$

$$\frac{\partial E}{\partial t} = i\omega A e^{i\theta} + \eta \Sigma'_n \, i n \omega c_n \, e^{in(\theta+\varphi)} + c.c. \qquad (5.14)$$

$$\frac{\partial^2 E}{\partial t^2} = - \omega^2 A e^{i\theta} - \eta \Sigma'_n \, n^2 \omega^2 c_n \, e^{in(\theta+\varphi)} + c.c. \qquad (5.15)$$

As far as the last term of (5.2) is concerned, we remark that, putting

$$E_o = A\, e^{i\theta} + A^* e^{-i\theta} = 2a \cos(\theta + \varphi) \,, \qquad (5.16)$$

so that E_o is the principal term, we have:

$$E = E_o + O(\eta) \qquad (5.17)$$

and, assuming F(E) differentiable,:

$$\eta\, F(E) = \eta\, F(E_o) + O(\eta^2) = \eta\, F(2a \cos(\theta + \varphi)) + O(\eta^2) \,.$$

Hence the function $F(E_o)$ is periodic of period 2π with respect to $\theta + \varphi$, so that it can be expanded in a Fourier series:

$$F(E_o) = \sum_o^\infty p_n \, e^{in(\theta+\varphi)} + c.c. \qquad (5.18)$$

where(3)

$$p_n = \frac{1}{2\pi} \int_{-\pi}^{\pi} F(2a \cos(\theta + \varphi)\, e^{-in(\theta + \varphi)} \, d(\theta + \varphi) \,. \qquad (5.19)$$

Notice also that p_n depends only on a, and consequently only on z, so that p_n does not depend on t. Hence:

$$\eta \frac{\partial^2 F(E)}{\partial t^2} = - \eta \omega^2 \sum_o^\infty n^2 p_n \, e^{in(\theta+\varphi)} + O(\eta^2) + c.c. \qquad (5.20)$$

Substituting (5.13), (5.14), (5.15), (5.20) into (5.2) we have:

---

(3) Strictly speaking for $n = 0$ we have 4π instead of 2π in the right-hand side of (5.19)

$$-\omega^2\epsilon\mu A e^{i\theta} - 2i\omega\sqrt{\epsilon\mu}\,\frac{dA}{dz}e^{i\theta} + \eta\,\Sigma'\Big(-n^2\omega^2\epsilon\mu c_n - 2i\omega n\sqrt{\epsilon\mu}\,\frac{dc_n}{dz} +$$

$$+\frac{d^2c_n}{dz^2}\Big)e^{in(\theta+\varphi)} + O(\eta^2) + \text{c.c.} = -\epsilon\mu\omega^2 A e^{i\theta} - \eta\epsilon\mu\,\Sigma' n^2\omega^2 c_n e^{in(\theta+\varphi)} -$$

$$-\mu\eta\omega^2\sum_0^\infty n^2 p_n e^{in(\theta+\varphi)} + \eta\mu\gamma_1 i\omega A e^{i\theta} + \eta\mu\gamma_1\Sigma' in\omega c_n e^{in(\theta+\varphi)} +$$

$$+ O(\eta^2) + \text{c.c.}$$

Equating the terms in $e^{i\theta}$ and $e^{in(\theta+\varphi)}$, after some simplifications, we obtain the following system:

$$\frac{dA}{dz} = -\eta\,\frac{i\omega\mu\gamma_1 A - \mu\omega^2 p_1 e^{i\varphi}}{2i\omega\sqrt{\epsilon\mu}}, \tag{5.21}$$

$$2i\omega n\sqrt{\epsilon\mu}\,\frac{dc_n}{dz} - \frac{d^2c_n}{dz^2} = \mu\omega^2 n^2 p_n - in\omega\mu\gamma_1 c_n, \tag{5.22}$$

with $n = 0, 2, 3, \ldots$.

By means of (5.21), (5.22) and suitable initial conditions to be discussed below we can solve for A and c_n, and hence E. Then by means of (2.3), (2.4) we can solve for H, so that (2.5) is satisfied up to the terms $O(\eta^2)$.

We shall come back to this result in Sect. 8. Now we limit ourselves to the remark that in what follows the terms $O(\eta^2)$ will be neglected without explicit mention. We notice, in addition, that differentiation and summation operations on the series have been often interchanged, which should be discussed even if convergence of the series takes place: we shall see below that in practice the series reduces to a finite sum.

It is worth noticing that (5.21) implies, recalling that $A = ae^{i\varphi}$ and p_1 depends only on a, that dA/dz is $\eta e^{i\varphi}$ times a function of a, in agreement with (5.9).

6. To test the validity of such an approximate method, suppose $F(E) = 0$ but $\gamma = \eta\gamma_1 \neq 0$. In this case the problem becomes linear, but with weak conductivity. Let us prove that the approximate method leads to the same expressions (of course up to terms of order η^2) for E and H obtained by solving exactly the Maxwell equations.

In the linear case if the source is sinusoidal of frequency ω, the same is true for the field, and thus all the c_n must vanish. Furthermore p_n, and in particular p_1, vanish since $F(E)$ is zero, and by (5.21) (recall that $\eta\gamma_1 = \gamma$) we than have:

$$\frac{dA}{dz} = -\frac{1}{2}\gamma\sqrt{\frac{\mu}{\epsilon}}\,A \tag{6.1}$$

whence (A_o being some constant):

$$A = A_o \exp\left(-\frac{\gamma}{2}\sqrt{\frac{\mu}{\epsilon}}\,z\right) \tag{6.2}$$

$$E = A_o \exp\left(-\frac{\gamma}{2}\sqrt{\frac{\mu}{\epsilon}}\,z\right) e^{i\theta}. \tag{6.3}$$

To compute H notice that by (5.12) and (6.1) we have:

$$\frac{\partial E}{\partial z} = -\left(i\omega\sqrt{\epsilon\mu} + \frac{\gamma}{2}\sqrt{\frac{\mu}{\epsilon}}\right) A e^{i\theta} + c.c. = -i\omega\sqrt{\epsilon\mu}\left(1 - \frac{i\gamma}{2\epsilon\omega}\right) A e^{i\theta} + c.c. \tag{6.4}$$

Hence by (2.4)

$$\frac{\partial H}{\partial t} = i\omega\sqrt{\frac{\epsilon}{\mu}}\left(1 - \frac{i\gamma}{2\epsilon\omega}\right) A e^{i\theta} + c.c. \tag{6.5}$$

Integrating over t and taking into account the sinusoidal dependence of H on t, we have:

$$H = \sqrt{\frac{\epsilon}{\mu}}\left(1 - \frac{i\gamma}{2\epsilon\omega}\right) A e^{i\theta} + c.c. \tag{6.6}$$

It is possible to verify that (6.6) satisfies (2.3) up to terms of order $\gamma^2 = \gamma_1^2\eta^2$, as is allowed within our approximation. Let us now compare (6.3) with the exact solution of (5.2) which in this case reduces to:

$$\frac{\partial^2 E}{\partial z^2} = \epsilon\mu\frac{\partial^2 E}{\partial t^2} + \mu\gamma\frac{\partial E}{\partial t}. \tag{6.7}$$

Since the field has to be sinusoidal, set:

$$E = \mathscr{E}_o\, e^{i\omega(t - kz)} + c.c. \tag{6.8}$$

where κ and $\mathscr{E}_o$ are some constants. We immediately have that (6.7) is satisfied if:

$$\kappa^2 = \epsilon\mu - i\frac{\gamma\mu}{\omega}, \qquad (6.9)$$

and furthermore, since for $z = 0$ both values of E, the exact and the approximate one, must coincide, we have:

$$\mathscr{E}_o = A_o .$$

On the other hand:

$$\kappa = \pm \sqrt{\epsilon\mu}\sqrt{\left(1 - \frac{i\gamma}{\omega\epsilon}\right)}.$$

Now, since $\frac{\gamma}{\epsilon\,\omega} = \eta\frac{\gamma_1}{\epsilon\,\omega}$, the above square root can be expanded in powers of $\gamma/\epsilon\omega$ neglecting the terms in η^2 which are $O(\eta^2)$, so that we get:

$$\kappa = \pm \sqrt{\epsilon\mu}\left(1 - \frac{i\gamma}{2\epsilon\omega}\right), \qquad (6.10)$$

where the positive sign has to be chosen to obtain wave propagation along the positive direction of the z axis. Then, substituting (6.10) into (6.8), it is easily seen that E as given by this formula coincides with the value given by (6.3).

To prove also that the approximate and exact values of the magnetic field coincide, it will be enough, by (2.4) which determines H, to show the coincidence, always within our approximation, of the values of $\partial E/\partial z$ obtained by (6.9) and (6.4).

As a matter of fact, differentiating (6.8) and recalling the expression (6.10) for κ, we have:

$$\frac{\partial E}{\partial z} = - i\omega\sqrt{\epsilon\mu}\left(1 - \frac{i\gamma}{2\epsilon\omega}\right) A e^{i\theta} + c.c.$$

which coincides with (6.4).

6.7 Let us come back to the nonlinear case, assuming for simplicity that $\gamma = 0$. Let us first establish some properties of p_n defined by (5.19).

Let us begin by remarking that the expression for p_n may be written in the following form:

$$p_n = \frac{1}{2\pi}\int_{-\pi}^{\pi} F(2a\cos(\theta+\varphi))\cos n(\theta+\varphi)\,d(\theta+\varphi) +$$

$$+ \frac{i}{2\pi}\int_{-\pi}^{\pi} F(2a\cos(\theta+\varphi))\sin n(\theta+\varphi)\,d(\theta+\varphi)\,. \tag{7.1}$$

The integrand is odd in the last integral of (7.1), so that the integral vanishes, and p_n is real for any n. In particular p_1 is real; hence by (5.21) $e^{-i\varphi}dA/dz$ is purely imaginary. Then (5.9) implies $da/dz = 0$, i.e. within our approximation a is equal to the constant value $a(0) = a_o$: the amplitude of the fundamental wave undergoes no variation with z. Furthermore, p_n is also constant, since it depends only on a which is constant.

In addition (5.21) and (5.9) yield:

$$ia\frac{d\varphi}{dz} = e^{-i\varphi}\frac{dA}{dz} = \eta\frac{\mu\omega^2 p_1}{2i\omega\sqrt{\epsilon\mu}} = -\eta\frac{\mu i\omega}{2\sqrt{\epsilon\mu}}\,p_1\,, \tag{7.2}$$

whence

$$\frac{d\varphi}{dz} = -\eta\frac{\omega\mu p_1}{2a\sqrt{\epsilon\mu}}\,,\quad \varphi = -\eta\frac{\omega\mu p_1}{2a\sqrt{\epsilon\mu}}\,z + \varphi_o \tag{7.3}$$

where φ_o is the value assumed by φ at $z = 0$.

Let us now proceed to solve (5.22), leading to the computation of c_n. To this end, we notice that through an integration, we get:

$$2in\omega\sqrt{\epsilon\mu}\,c_n - \frac{dc_n}{dz} = h_n + n^2\omega^2\mu p_n z\,, \tag{7.4}$$

where h_n denotes a constant.

Now a particular solution of (7.4) is:

$$\bar{c}_n = \frac{n^2\omega^2\mu p_n}{2i\omega n\sqrt{\epsilon\mu}}\,z + \kappa_n = -\frac{ni\omega}{2}\sqrt{\frac{\mu}{\epsilon}}\,p_n z + \kappa_n \tag{7.5}$$

where κ_n is another constant that may be determined easily: indeed, substituting (7.5) into (7.4) we have:

$$n^2\omega^2\mu p_n z + 2in\omega\sqrt{\epsilon\mu}\,\kappa_n - \frac{n^2\omega^2\mu p_n}{2i\omega n\sqrt{\epsilon\mu}} = h_n + n^2\omega^2\mu\, p_n z\,.$$

Choosing κ_n in such a way as to satisfy this relation, it is immediately seen that (7.5) solves (7.4).

Therefore the general solution of (7.4) is (ℓ_n is some other constant)

$$c_n = \ell_n\, e^{2in\omega\sqrt{\epsilon\mu}\,z} + \bar{c}_n\,. \tag{7.6}$$

Hence the harmonic of E of frequency $n\omega$ is, by (7.6), (7.5) and (7.3):

$$\eta c_n e^{in(\theta+\varphi)} = \eta\,\ell_n \exp\left\{in\omega\left(t + \sqrt{\epsilon\mu}\,z - \frac{\eta p_1 z}{2a\sqrt{\epsilon\mu}} + \varphi_o\right)\right\} + \eta\bar{c}_n e^{in(\theta+\varphi)}. \tag{7.7}$$

Now, since η is very small, the coefficient of z in the first exponential in the right-hand side of (7.7) is positive. Thus the exponential represents a wave propagating along the negative direction of the z axis; since such a wave has been excluded we must have $\ell_n = 0$ and $c_n = \bar{c}_n$.

Let us now proceed to the computation of the magnetic field H. By (2.4) and (5.12) we have:

$$\frac{\partial H}{\partial t} = -\frac{1}{\mu}\frac{\partial E}{\partial z} = i\omega\sqrt{\frac{\epsilon}{\mu}}\,Ae^{i\theta} - \frac{1}{\mu}\frac{dA}{dz}e^{i\theta} + \frac{\eta}{\mu}\,\Sigma'_n\left(in\omega\sqrt{\epsilon\mu}\,c_n - \frac{dc_n}{dz}\right)e^{in(\theta+\varphi)} + \text{c.c.} \tag{7.8}$$

Hence, integrating over t and neglecting the static field which would arise from such an integration and would superpose with the time-variable field (the only one of interest to us), we have, recalling that $d\theta/dt = i\omega$:

$$H = \sqrt{\frac{\epsilon}{\mu}}\,Ae^{i\theta} - \frac{1}{\mu i\omega}\frac{dA}{dz}e^{i\theta} + \eta\,\Sigma'\left(\sqrt{\frac{\epsilon}{\mu}}\,c_n - \frac{1}{\mu i\omega n}\frac{dc_n}{dz}\right)e^{in(\theta+\varphi)} + \text{c.c.} \tag{7.9}$$

It is worth remarking that the expression (7.9) for H also satisfies the other Maxwell equation (2.3) which now becomes, in view of (5.1) and the fact that $\gamma = 0$:

$$- \frac{\partial H}{\partial z} = \epsilon \frac{\partial E}{\partial t} + \frac{\partial}{\partial t} \eta F(E). \tag{7.10}$$

Now, neglecting the terms $O(\eta^2)$, (7.9) yields:

$$- \frac{\partial H}{\partial z} = \epsilon i \omega A e^{i\theta} - 2\sqrt{\frac{\epsilon}{\mu}} \frac{dA}{dz} e^{i\theta} - \eta \Sigma' \frac{1}{\mu i \omega n} \left(2 i n \sqrt{\epsilon\mu}\, \omega \frac{dc_n}{dz} - \frac{d^2 c_n}{dz^2} - \right.$$

$$\left. - i^2 n^2 \omega^2 \epsilon \mu c_n\right) e^{in(\theta+\varphi)} + \text{c.c.}$$

and recalling (5.21) and (5.22) we have:

$$- \frac{\partial H}{\partial z} = \epsilon i \omega A e^{i\theta} + \eta i \omega p_1 e^{i(\theta+\varphi)} - \eta \Sigma' \frac{1}{\mu i \omega n} (\mu \omega^2 n^2 p_n - i^2 n^2 \omega^2 \epsilon \mu c_n) e^{in(\theta+\varphi)} +$$

$$+ \text{c.c.} = \epsilon (i \omega A e^{i\theta} + \eta \Sigma' i n \omega c_n e^{in(\theta+\varphi)}) + \eta \sum_1^\infty i \omega n p_n e^{in(\theta+\varphi)} + \text{c.c.}$$

so that, by (5.14), taking the time derivative of (5.18), we obtain (7.10).

Let us notice that, recalling (5.21) and (5.22), we have:

$$H = \sqrt{\frac{\epsilon}{\mu}} A e^{i\theta} + \eta \frac{1}{2\sqrt{\epsilon\mu}} p_1 e^{i(\theta+\varphi)} + \eta \Sigma' \left(\sqrt{\frac{\epsilon}{\mu}} c_n - \frac{1}{\mu i \omega n} \frac{dc_n}{dz}\right) e^{in(\theta+\varphi)} + \text{c.c.} \tag{7.11}$$

Let us now proceed to the computation of the constants coming from the integration of (5.21) and (5.22).

To this end, some conditions on the plane $z = 0$ have to be specified. For example, let the electric field be assigned as a sinusoidal function of the time: this is the most usual assumption in nonlinear Optics. That is, for $z = 0$ let $E = \mathscr{E}_o e^{i\omega t} + \text{c.c.}$[1]

Then, recalling (5.6), (7.3) and (7.5) we have $\mathscr{E}_o = a_o$, $\varphi_o = 0$, $c_n(0) = \overline{c}_n(0) = 0$, whence $\kappa_n = 0$ so that:

$$E = a_o e^{i(\theta+\varphi)} - \eta \sum_2^\infty \frac{i n \omega}{2} \sqrt{\frac{\mu}{\epsilon}} p_n z e^{in(\theta+\varphi)}. \tag{7.12}$$

The summation goes from 2 to ∞ because the $n = 0$ term vanishes and φ is expressed by (7.3). In addition we have:

$$H = \sqrt{\frac{\epsilon}{\mu}} a_o e^{i(\theta+\varphi)} + \eta \frac{p_1}{2\sqrt{\epsilon\mu}} e^{i(\theta+\varphi)} - \eta \Sigma' \left(\frac{i n \omega}{2} p_n z - \frac{p_n}{2\sqrt{\epsilon\mu}}\right) e^{in(\theta+\varphi)}. \tag{7.13}$$

---

(1) This hypothesis amounts to taking $\varphi_o = 0$. We can always reduce the problem to this case by changing the time origin.

Let us now come to the case in which the plane $z = 0$ separates two different media; more precisely for $z < 0$ there is the medium already considered in Sect. 3. Let the incident field E be sinusoidal, specified by:

$$E_i(0,t) = \mathscr{E}_o e^{i\omega t} + c.c. \tag{7.14}$$

We thus have from (3.5) and (3.3), in which we substitute (5.6) and (7.11):

$$\sqrt{\frac{\epsilon_o}{\mu}}\left[a_o e^{i(\omega t+\varphi_o)} + \eta\, \Sigma' \kappa_n e^{in(\omega t+\varphi_o)}\right] + \sqrt{\frac{\epsilon}{\mu}}\, a_o e^{i(\omega t+\varphi_o)} +$$

$$+ \eta \frac{p_1}{2\sqrt{\epsilon\mu}} e^{i(\omega t+\varphi_o)} + \eta\, \Sigma'\left[\sqrt{\frac{\epsilon}{\mu}}\,\kappa_n + \frac{1}{2}\frac{p_n}{\sqrt{\epsilon\mu}}\right] e^{in(\omega t+\varphi_o)} = 2\sqrt{\frac{\epsilon_o}{\mu}}\,\mathscr{E}_o e^{i\omega t}. \tag{7.15}$$

Comparing the coefficients of the terms in $e^{i\omega t}$ we have:

$$\left[\sqrt{\frac{\epsilon_o}{\mu}} + \sqrt{\frac{\epsilon}{\mu}}\right] a_o e^{i\varphi_o} + \frac{\eta p_1}{2\sqrt{\epsilon\mu}} e^{i\varphi_o} = 2\sqrt{\frac{\epsilon_o}{\mu}}\,\mathscr{E}_o . \tag{7.16}$$

Hence we get:

$$\varphi_o = 0; \qquad a_o = \frac{2\sqrt{\frac{\epsilon_o}{\mu}}\,\mathscr{E}_o}{\sqrt{\frac{\epsilon_o}{\mu}} + \sqrt{\frac{\epsilon}{\mu}} + \frac{\eta p_1}{2\sqrt{\epsilon\mu}\, a_o}} \tag{7.17}$$

Comparing also the coefficients of the terms in $e^{in\omega t}$ we have:

$$\left[\sqrt{\frac{\epsilon_o}{\mu}} + \sqrt{\frac{\epsilon}{\mu}}\right]\kappa_n = -\frac{p_n}{2\sqrt{\epsilon\mu}} \;; \quad \kappa_n = \frac{-p_n}{2\sqrt{\epsilon\mu}\left[\sqrt{\frac{\epsilon_o}{\mu}} + \sqrt{\frac{\epsilon}{\mu}}\right]} . \tag{7.18}$$

Hence we get:

$$E = a_o e^{i(\theta+\varphi)} - \eta\, \Sigma'\left[\frac{p_n}{2(\sqrt{\epsilon_o\epsilon} + \epsilon)} + \frac{n i\omega}{2}\sqrt{\frac{\mu}{\epsilon}}\, p_n z\right] e^{in(\theta+\varphi)} + c.c. \tag{7.19}$$

$$H = \left[\sqrt{\frac{\epsilon}{\mu}}\, a_o + \frac{\eta p_1}{2\sqrt{\epsilon\mu}}\right] e^{i(\theta+\varphi)} - \eta\, \Sigma'\left[\frac{p_n}{2\sqrt{\mu}(\sqrt{\epsilon_o} + \sqrt{\epsilon})} + \frac{ni\omega}{2} p_n z - \frac{p_n}{\sqrt{\epsilon\mu}}\right] e^{in(\theta+\varphi)} + c.c. \tag{7.20}$$

Let us now find the field of the reflected wave on the plane $z = 0$. By (3.1) we have:

$$E_r(0,t) = E(0,t) - E_i(0,t) . \tag{7.21}$$

Hence:

$$E_r(0,t) = - \frac{\sqrt{\epsilon} - \sqrt{\epsilon_o} + \frac{\eta p_1}{2\sqrt{\epsilon}\, a_o}}{\sqrt{\epsilon_o} + \sqrt{\epsilon} + \frac{\eta p_1}{2\sqrt{\epsilon}\, a_o}} \mathscr{E}_o e^{i\omega t} - \eta \Sigma' \frac{p_n}{2(\sqrt{\epsilon_o \epsilon} + \epsilon)} e^{in\omega t} + c.c. \tag{7.22}$$

$E_r(z,t)$ is obtained by replacing ωt by $\omega t + \sqrt{\epsilon \mu}\, z$ in the argument of the exponential.

6.8 Let us now discuss the results obtained also from the physical standpoint. To begin with, we have that in a nonlinear medium, even if on the surface there is only a wave of frequency ω, or the source is sinusoidal of frequency ω, there is propagation, beyond the principal wave, also of waves of frequency $n\omega$, at least in correspondence to those values of n for which $p_n \neq 0$. These waves have been called n-th order waves. Furthermore, also in the reflected wave there occur n-th order harmonic terms, i.e. terms of frequency $n\omega$. This means that the nonlinear medium acts as a multiplier of optical frequencies: the theory so far described allows the computation of the intensity of the n-th order harmonic terms, both transmitted and reflected, and the determination of the dependence of the intensity on the amplitude a_o of the principal wave.

Notice also that if $p_1 \neq 0$ the nonlinearity changes the phase propagation velocity of quantities depending on the amplitude. Indeed we have, recalling (7.3) and supposing again $\varphi_o = 0$:

$$n(\theta + \varphi) = n\omega[t - (\sqrt{\epsilon\mu} + \eta \frac{p_1 \mu}{2 a_o \sqrt{\epsilon \mu}})z] . \tag{8.1}$$

Hence the phase velocity is:

$$v_f = \left\{ \sqrt{\epsilon \mu} + \frac{\eta p_1 \mu}{2 a_o \sqrt{\epsilon \mu}} \right\}^{-1} , \tag{8.2}$$

so that v_f is the same for the principal wave as for the higher order harmonic terms. Hence the refractive index n corresponding to the medium $z < 0$ is given by:

$$n = \frac{1}{v_f \sqrt{\epsilon_o \mu}} = \sqrt{\frac{\epsilon}{\epsilon_o}} + \frac{\eta p_1}{2 a_o \sqrt{\epsilon \epsilon_o}} . \tag{8.3}$$

Hence, dividing (7.17) by $\sqrt{\frac{\epsilon_o}{\mu}}$, we get:

$$a_o = \frac{2 \mathscr{E}_o}{1 + n} \tag{8.4}$$

in agreement with the well known formula for refraction with normal incidence.

Let us now compute p_n for some particular cases. Notice that, putting for short $\psi = \theta + \varphi$, (7.1) can be rewritten as:

$$p_n = \frac{1}{2\pi} \int_0^{\pi} F(2a \cos \psi) \cos n\psi \, d\psi + \frac{1}{2\pi} \int_{-\pi}^{0} F(2a \cos \psi) \cos n\psi \, d\psi . \tag{8.5}$$

Replacing ψ by $\psi + \pi$ in the second integral, we obtain:

$$p_n = \frac{1}{2\pi} \int_0^{\pi} F(2a \cos \psi) \cos n\psi \, d\psi + \frac{1}{2\pi} \int_0^{\pi} F(-2a \cos \psi) \cos n\psi \cos n\pi \, d\psi , \tag{8.5'}$$

so that we have:

$$F(E) = F(-E), \qquad p_n = 0 \qquad \text{for all odd } n \tag{8.6}$$

$$F(E) = -F(-E), \qquad p_n = 0 \qquad \text{for all even } n . \tag{8.6'}$$

Hence, in the case of (8.6) there are higher order waves only of even order, and in the other one only waves of odd order. Furthermore, for the case of (8.6), $p_1 = 0$ and assuming $\varphi_o = 0$ we have $\varphi = 0$, so that the nonlinearity does not change the wave propagation velocity which remains $1/\sqrt{\epsilon \mu}$.

For example let:

$$F(E) = \alpha E^2 \tag{8.7}$$

with α a constant. Then we have:

$$F(2a \cos \psi) = 4 \alpha a^2 \cos^2 \psi = 2 \alpha a^2 (1 + \cos 2\psi).$$

In this case only p_o and p_2 do not vanish. The term in p_o may give rise to a static field of no interest to us.

We have instead:

$$p_2 = \frac{2\alpha a^2}{2\pi} \int_{-\pi}^{\pi} \cos^2 2\psi \, d\psi = \alpha a^2, \tag{8.8}$$

so that in the nonlinear medium there is propagation of a field of frequency 2ω, whose amplitude can easily be computed by means of (7.19) and is in any case proportional to the square of the principal wave. Therefore illuminating a nonlinear plate by means of a laser wave, whose frequency corresponds to a wave length $\lambda = 6940\ A^o$, it is possible to obtain also a wave with $\lambda = 3470\ A^o$. That is, from laser heating can be obtained through a nonlinear medium a violet light wave. This result is confirmed by experiments.

Let us now turn to the case in which (β is constant):

$$F(E) = \beta E^3 \tag{8.9}$$

Then:

$$F(2a\cos\psi) = 8\beta a^3 \cos^3\psi = 4\beta a^3 \cos\psi\,(1 + \cos 2\psi) = 4\beta a^3 (\cos\psi + \frac{1}{2}\cos\psi +$$
$$+ \frac{1}{2}\cos 3\psi) = 2\beta a^3 (3\cos\psi + \cos 3\psi).$$

In this case only p_1 and p_3 are different from zero. We have:

$$p_1 = 3\beta a^3, \qquad p_3 = \beta a^3.$$

Then by (8.2) the velocity of the waves is changed by a quantity proportional to a^2, that is, it is proportional to the square of the principal wave amplitude, whereas there is a wave of frequency 3ω proportional to the third power of the same amplitude.

Before closing the section, let us return to the approximation method applied here as well as in the preceding sections. We have seen that the approximation solutions satisfy the conditions assigned at $z = 0$ and the equation (5.2) up to terms $O(\eta^2)$. Therefore it is reasonable to presume that, as happens for ordinary differential equations, the difference between the exact solution and the approximate solution is of order $O(\eta^2)$, which can also be neglected with respect to the terms in η occurring in the expressions for E and H as long as z is not too large. This condition is certainly satisfied when, as happens in applications, we consider only plates of finite thickness, or plates such as those discussed in Sect. 2. In this case, however, we should also take into account the wave reflected from the face $z = s$ of the plate, i.e. of a wave propagating along the negative direction of the z axis.

Hence it is probable that the above results will hold provided the contribution of the reflected wave can be neglected. Anyway, a deeper investigation of the approximate methods applied to the propagation within a plate of finite thickness would be in order, even if it leads to complicated calculations.

In the next section we shall confirm by a different way the results so far obtained, always neglecting the wave reflected on the plane $z = s$.

6.9 Consider again the solution (4.2), in which we put $z_o = \alpha = 0$; assume in addition the validity of (5.1), so that:

$$p(E) = \sqrt{\epsilon\mu + \eta\mu \frac{dF(E)}{dE}} \tag{9.1}$$

Then, taking a_o to be real, we get from (4.2):

$$G(u) = a_o e^{i\omega u} + c.c. \tag{9.2}$$

We have:

$$E(z,t) = a_o \exp\{i\omega(t - \sqrt{\epsilon\mu + \eta\mu \frac{dF}{dE}}\, z)\} + c.c. \tag{9.3}$$

Of course, $E(0,t) = a_o e^{i\omega t} + c.c.$ coincides with the electric field at $z = 0$ which is known by assumption, as in the first example of Sect. 7. Hence we can prove that, up to terms of order η^2, $E(z,t)$ (for short E) as given by (9.3) coincides with (7.12). To this end notice that, for z and t fixed, E is a function of η. We will write:

$$E = E(\eta). \tag{9.4}$$

Now, expanding $E(\eta)$ in a Taylor series, and negelecting the terms of order higher than one, we get:

$$E = E_o' + \eta(\frac{\partial E}{\partial \eta})_o + O(\eta^2), \tag{9.5}$$

where

$$E_o' = a_o e^{i\omega(t - \sqrt{\epsilon\mu} z)} + c.c. = 2 a_o \cos\theta, \tag{9.6}$$

which coincides with (5.16), in which we put $a = a_o$, $\varphi = 0$, and $(\partial E/\partial\eta)_o$ is the partial derivative of E with respect to η at $\eta = 0$.

Now denote by $(\partial F(E)/\partial E)_o$ the derivative of F(E) with respect to E at $\eta = 0$. Since by (9.5) we have $(\partial E_o'/\partial E)_o = 1$, it follows that:

$$(\frac{\partial F(E)}{\partial E})_o = (\frac{\partial F(E)}{\partial E_o'})_o = \frac{\partial F(E_o')}{\partial E_o'} . \tag{9.7}$$

Now compute $\eta(\partial E/\partial\eta)_o$ by differentiating (9.3): recalling also (9.5') we have:

$$\eta(\frac{\partial E}{\partial \eta})_o = - \eta a_o i\omega \frac{e^{i\omega(t - \sqrt{\epsilon\mu} z)}}{2\sqrt{\epsilon\mu}} \mu (\frac{\partial F}{\partial E})_o z + c.c. = - \eta a_o i\omega z \frac{e^{i\omega(t - \sqrt{\epsilon\mu} z)}}{2\sqrt{\epsilon\mu}} .$$

$$\mu \frac{\partial F(E_o')}{\partial E_o'} + c.c. = - \frac{\eta}{2}\sqrt{\frac{\mu}{\epsilon}} z \frac{\partial E_o'}{\partial t} \frac{\partial F(E_o')}{\partial E_o'} = - \frac{\eta}{2}\sqrt{\frac{\mu}{\epsilon}} z \frac{\partial F(E_o')}{\partial t} . \tag{9.8}$$

Now recalling that E_o' coincides with E_o of (5.16) provided we set $\varphi = 0$ in E, by (5.18) we have:

$$\eta(\frac{\partial E}{\partial \eta})_o = - \frac{\eta}{2}\sqrt{\frac{\mu}{\epsilon}} z \sum_1^\infty i\omega n\, p_n e^{in\theta} + c.c. \tag{9.9}$$

Hence, substituting into (9.5), we get:

$$E = a_o e^{i\theta} - \frac{\eta}{2}\sqrt{\frac{\mu}{\epsilon}} z i\omega p_1 e^{i\theta} - \frac{\eta}{2}\sqrt{\frac{\mu}{\epsilon}} z \sum_2^\infty i\omega n\, p_n e^{in\theta} + c.c. \tag{9.10}$$

and this formula coincides with (7.12) up to terms $O(\eta^2)$. Notice indeed that now $\varphi_o = 0$, and hence by (7.3) for z not too large $\varphi = O(\eta)$. Hence we have:

$$a_o e^{i(\theta + \varphi)} = a_o e^{i\theta}(1 + i\varphi) + O(\eta^2) = a_o e^{i\theta}(1 - \frac{\eta i\omega\mu}{2 a_o \sqrt{\epsilon\mu}} p_1 z) + O(\eta^2). \tag{9.11}$$

Hence the first term of (7.12) coincides with the sum of the first two terms of (9.10). Furthermore, since $\varphi = O(\eta)$, we have $\eta\, e^{in(\theta+\varphi)} = \eta\, e^{in\theta} + O(\eta^2)$, i.e. the two sums also coincide. The two expressions (7.12) and (9.10) for E are thus the same, as was claimed.

Let us now turn to the computation of the magnetic field, always in the approximation $O(\eta^2)$. Let us apply formula (4.10). Since by (9.1) we have, expanding the square root in a power series of η:

$$p(E) = \sqrt{\epsilon\mu + \eta\,\mu\,\frac{\partial F(E)}{\partial E}} = \sqrt{\epsilon\mu} + \frac{\eta}{2}\sqrt{\frac{\mu}{\epsilon}}\,\frac{\partial F(E)}{\partial E} + O(\eta^2) \qquad (9.12)$$

by (4.10), assuming the integration constants to be zero which represents a static field, we get:

$$H = \sqrt{\frac{\epsilon}{\mu}}\,E + \frac{\eta}{2}\,\frac{1}{\sqrt{\epsilon\mu}}\,F(E) + O(\eta^2). \qquad (9.13)$$

Now, up to terms $O(\eta^2)$, $\eta F(E) = \eta F(E_o)$ given by (5.18). Hence, recalling (7.12) and that up to terms $O(\eta)$, $\eta\, e^{i(\theta+\varphi)} = e^{i\theta}$, we have:

$$H = \sqrt{\frac{\epsilon}{\mu}}\,a_o\,e^{i(\theta+\varphi)} - \frac{\eta}{2}\sum_{2}^{\infty} in\omega p_n\, z\, e^{in(\theta+\varphi)} + \frac{\eta}{2\sqrt{\epsilon\mu}}\,p_1\,e^{i(\theta+\varphi)} +$$

$$+\,\eta \sum_{o}^{\infty}{}' \frac{p_n}{2\sqrt{\epsilon\mu}}\,e^{in(\theta+\varphi)} + O(\eta^2), \qquad (9.14)$$

which coincides with (7.13), provided the static field is, as usual, neglected.

We observe that this method yields more rapidly and rigorously the results obtained through the approximate method of the preceding sections. However, it is easier to extend the above method to the case of an absorbing medium, also of finite thickness, and it is for this reason that we have found it convenient to describe it in some detail.

6.10 In the treatment of the above sections we have assumed the medium to be non-linear with no dispersion, so that the dielectric constant and the magnetic permeability are independent of frequency. The media occurring in non-linear Optics are, however, dispersive, so that ϵ is a function of the frequency, while μ can be considered constant, equal to the vacuum permeability.

Let us see how the above theory can be modified to account for the dispersion. We will not attempt, for brevity, to give a complete mathematical treatment, but rather we limit ourselves to some arguments which will be sufficient to obtain the results we are interested in.

To this end, consider again (5.6) where ϵ is to be replaced by $\epsilon(\omega)$ (ϵ_1 for short) in the principal term. In those terms of the series representing waves of frequency $n\omega$, corresponding to the dielectric constant $\epsilon_n = \epsilon(n\omega)$, θ has to be replaced by:

$$\theta_n = \omega(t - \sqrt{\epsilon_n \mu}\, z). \tag{10.1}$$

Of course, $\theta_1 = \omega(t - \sqrt{\epsilon_1 \mu}\, z)$.

Hence we have:

$$E = A e^{i\theta_1} + \eta \sum{}' c_n e^{in(\theta_n + \varphi)} + c.c. \tag{10.2}$$

Equations (5.7), (5.8) and (5.9) are still valid; so that repeating the arguments of Sect. 5 (taking into account the fact that in (5.2) ϵ must be replaced by ϵ_n in the coefficient of $\partial^2 E/\partial t^2$ corresponding to the term $c_n e^{in\theta_n}$), we find that the values of A, that is the values of a and φ, are left unchanged. On the contrary, since the term in c_n in the series is multiplied by $e^{in(\theta_n + \varphi)}$, while the corresponding term of the expansion of F(E) is always given by $\eta n^2 \omega^2 \mu p_n e^{in(\theta+\varphi)}$, (letting $\theta_1 = \theta$ for simplicity) we have the relation:

$$e^{in\theta_n} \left(2\, in\omega \sqrt{\epsilon_n \mu}\, \frac{dc_n}{dz} - \frac{d^2 c_n}{dz^2}\right) = n^2 \omega^2 p_n \mu e^{in\theta}, \tag{10.3}$$

whence:

$$2\, ni\omega \sqrt{\epsilon_n \mu}\, \frac{dc_n}{dz} - \frac{d^2 c_n}{dz^2} = n^2 \omega^2 p_n \mu e^{in(\theta - \theta_n)}. \tag{10.4}$$

Now we have:

$$in(\theta - \theta_n) = in\omega t - in\omega\sqrt{\epsilon_1 \mu}\, z + in\omega \sqrt{\epsilon_n \mu}\, z - in\omega t = - in\omega(\sqrt{\epsilon_1 \mu} - \sqrt{\epsilon_n \mu})z = - i\Delta_n z, \tag{10.5}$$

with

$$\Delta_n = n\omega(\sqrt{\epsilon_1\mu} - \sqrt{\epsilon_n\mu}). \tag{10.6}$$

Substituting into (10.4), after integration we get:

$$2n\omega i\sqrt{\epsilon_n\mu}\, c_n - \frac{dc_n}{dz} = \frac{\mu n^2 \omega^2 p_n}{-i\Delta_n} e^{-i\Delta_n z} + \kappa_n . \tag{10.7}$$

In order to determine a particular integral of (10.7), set:

$$\overline{c}_n = \frac{\kappa_n}{2n\omega i\sqrt{\epsilon_n\mu}} + h e^{-i\Delta_n z} . \tag{10.8}$$

The constant h is to be determined in such a way as to satisfy (10.7), so that h must satisfy the equation:

$$(2n\omega i\sqrt{\epsilon_n\mu} + i\Delta_n)h = -\frac{n^2\omega^2}{i\Delta_n}\mu p_n \tag{10.9}$$

whence:

$$h = -\frac{n^2\omega^2 p_n \mu}{i\Delta_n(2n\omega i\sqrt{\epsilon_n\mu} + i\Delta_n)} . \tag{10.10}$$

We have now to add this integral to the general integral of (10.7) with $\kappa_n = 0$ and $p_n = 0$; however, as has been seen in Sect. 7, in this way waves would be added that propagate along the negative direction of the z axis, which is a priori excluded. Hence $c_n = \overline{c}_n$. Since we limit ourselves to the case already examined in Sect. 7, in which for $z = 0$ E reduces to a sinusoidal function of frequency ω, we have $c_n(0) = 0$.

This condition is satisfied by putting:

$$\kappa_n = -2ni\omega\sqrt{\epsilon_n\mu}\, h.$$

To sum up we have, on account of (10.6) and (10.10):

$$c_n = \frac{p_n \mu n\omega}{\Delta_n(\sqrt{\epsilon_1\mu} + \sqrt{\epsilon_n\mu})} (e^{-i\Delta_n z} - 1). \tag{10.11}$$

Now,

$$|e^{-i\Delta_n z} - 1| = |(\cos \Delta_n z - 1) + \sin \Delta_n z| = \sqrt{2 - 2\cos \Delta_n z} = 2 \sin \frac{\Delta_n z}{2} . \tag{10.12}$$

Hence,

$$|c_n| = 2 \left| \frac{p_n \mu n \omega}{\sqrt{\epsilon_1 \mu} + \sqrt{\epsilon_n \mu}} \right| \left| \frac{\sin \Delta_n z/2}{\Delta_n z} \right| z. \tag{10.13}$$

For $\Delta_n z \to 0$, c_n, the n-th harmonic amplitude, reduces of course to (7.12). If $\Delta_n z \neq 0$, since ϵ_1 and ϵ_n are close to each other, we can state that c_n has more or less the same value as in the non-dispersive medium, except for the factor $2 |\sin (\Delta_n z/2)/\Delta_n z|$. This factor is 1 for $z = 0$ and becomes very small for $z \gg 1/\Delta_n$, so that the dispersivity weakens the higher order harmonics.

Actually, when detecting the second harmonic term (the only one existing, as already remarked, if the source is a laser, within the visible spectrum) we have to make $\Delta_2 = 0$ by means of suitable devices; in other words we try to make the system matched into the second harmonic term.

Let us finally remark that for the principal wave $\Delta = 0$, and hence for principal waves the system is always matched.

6.11 We have so far supposed the occurrence of a single wave of frequency ω in the nonlinear medium; let us now examine the case in which there are two waves of frequency ω_1 and ω_2, the amplitude of the first wave being much greater than that of the second one.

For simplicity take $F(E) = \alpha E^2$; that is assume a quadratic nonlinearity. A simple computation, sketched below, shows that the nonlinearity generates waves of frequency $2\omega_1$, $2\omega_2$, $\omega_1 + \omega_2$, $\omega_1 - \omega_2$, which are however very weak if the system is not matched(1) on them.

In this case, the situation will be studied in which the system is matched on the wave of frequency $\omega_1 + \omega_2$; this wave may interact with the remaining two waves

---

(1) The system is not matched on the frequencies $\Omega = \omega_1 \pm \omega_2$ when $\omega_1 \sqrt{\epsilon(\omega_1)\mu} \pm \omega_2 \sqrt{\epsilon(\omega_2)\mu} \neq \Omega \sqrt{\epsilon(\Omega)\mu}$. Proceeding as in Sect. 10 we can show that those waves are very weak.

so that we shall assume the occurrence in the medium of three waves, of frequency ω_1, ω_2 and $\omega_3 = \omega_1 + \omega_2$, respectively, corresponding to the dielectric constants $\epsilon_1, \epsilon_2, \epsilon_3$. Since the system is matched on the frequency ω_3 it will be such that:

$$\omega_3 \sqrt{\epsilon_3 \mu} = \omega_1 \sqrt{\epsilon_1 \mu} + \omega_2 \sqrt{\epsilon_2 \mu} . \tag{11.1}$$

As in the former sections, consider the principal waves corresponding to the three frequencies, given by:

$$A_1(z)e^{i\theta_1}, \quad A_2(z)e^{i\theta_2}, \quad A_3(z)e^{i\theta_3}, \tag{11.2}$$

where $\theta_1, \theta_2, \theta_3$ are specified by (10.1) with $\epsilon_n = \epsilon(\omega_n)$. In the present case we have (A^* being the complex conjugate of A):

$$F(E) = \alpha(A_1 e^{i\theta_1} + A_2 e^{i\theta_2} + A_3 e^{i\theta_3} + A_1^* e^{-i\theta_1} + A_2^* e^{-i\theta_2} + A_3^* e^{-i\theta_3})^2 . \tag{11.3}$$

Without entering into detailed computations, it is enough to remark that each term contains an exponential of $i(\theta_i \pm \theta_j)$, $i = j = 1, 2, 3$, of frequency $\omega_i \pm \omega_j$, $i=j=1,2,3$.

However the system is matched, by assumption, only on the wave of frequency $\omega_3 = \omega_1 + \omega_2$, in addition, of course, to those of frequency ω_1 and ω_2 (in agreement with the remark at the end of Sect. 10). Therefore, expanding the square we will take into account only the terms $\omega_1 + \omega_2 = \omega_3$, $\omega_3 - \omega_2 = \omega_1$, $\omega_3 - \omega_1 = \omega_2$ (that is the terms in θ_3, θ_1 and θ_2), because the remaining ones, being unmatched, can be neglected.

We thus obtain:

$$F(E) = 2\alpha(A_1 A_2 e^{i\theta_3} + A_1 A_3^* e^{-i\theta_2} + A_2 A_3^* e^{-i\theta_1} + \text{c.c.}). \tag{11.4}$$

Then, substituting in (5.2) and taking into account the fact that we have to set $\epsilon_1, \epsilon_2, \epsilon_3$ in correspondence to the terms of frequency $\omega_1, \omega_2, \omega_3$, we have:

$$\begin{aligned} &-2i\omega_1 \sqrt{\epsilon_1 \mu}\, \frac{dA_1}{dz} e^{i\theta_1} - 2i\omega_2 \sqrt{\epsilon_2 \mu}\, \frac{dA_2}{dz} e^{i\theta_2} - 2i\omega_3 \sqrt{\epsilon_3 \mu}\, \frac{dA_3}{dz} e^{i\theta_3} + \text{c.c.} = \\ &= -2\eta\mu\alpha(\omega_1^2 A_2^* A_3 e^{i\theta_1} + \omega_2^2 A_1^* A_3 e^{i\theta_2} + \omega_3^2 A_1 A_2 e^{i\theta_3} + \text{c.c.}) \end{aligned} \tag{11.5}$$

Whence:

$$\frac{dA_1}{dz} = -i\,\frac{\eta\mu\alpha\omega_1}{\sqrt{\epsilon_1\mu}}\,A_2^* A_3 = -i\kappa_1\eta A_2^* A_3\,, \tag{11.6}$$

$$\frac{dA_2}{dz} = -i\,\frac{\eta\mu\alpha\omega_2}{\sqrt{\epsilon_2\mu}}\,A_1^* A_3 = -i\kappa_2\eta A_1^* A_3\,, \tag{11.7}$$

$$\frac{dA_3}{dz} = -i\,\frac{\eta\mu\alpha\omega_3}{\sqrt{\epsilon_3\mu}}\,A_1 A_2 = -i\kappa_3\eta A_1 A_2\,, \tag{11.8}$$

the meaning of the real numbers $\kappa_1, \kappa_2, \kappa_3$ being obvious.

Now, since $|A_1|$ is, by assumption, much greater than $|A_2|$ and $|A_3|$, the increase of A_1 due to the nonlinearity during the propagation can be neglected; in other words, A_1 will be assumed constant. Since nothing prevents us from taking $A_1(0)$ real and positive, A_1 will be real and equal to $A_1(0)$.

Then to solve (11.7) and (11.8) let us look for two particular integrals putting:

$$A_2 = u_2 e^{i\Gamma z}\,, \qquad A_3 = u_3 e^{i\Gamma z}\,, \tag{11.9}$$

u_1, u_2, Γ being constants. We have:

$$i\,\Gamma u_2 = -i\eta\kappa_2 A_1 u_3\,, \qquad i\,\Gamma u_3 = -i\eta\kappa_3 A_1 u_2\,. \tag{11.10}$$

Hence:

$$\Gamma^2 = \eta^2 \kappa_2\kappa_3\,|A_1|^2\,. \tag{11.11}$$

Since $\kappa_2\kappa_3 \geq 0$, (11.11) implies that Γ can assume only the following values:

$$\Gamma' = \eta\sqrt{\kappa_2\kappa_3}\,|A_1|\,, \qquad \Gamma'' = -\eta\sqrt{\kappa_2\kappa_3}\,|A_1| = -\Gamma'\,.$$

Then, denoting by u_2' and u_2'' the values of u_2 corresponding to Γ' and Γ'' we have:

$$A_2 = u_2' e^{i\Gamma' z} + u_2'' e^{i\Gamma'' z}\,, \tag{11.12}$$

so that from (11.10) we obtain:

$$u'_3 = -\eta\frac{\kappa_3 |A_1|}{\Gamma'} u'_2 = -\sqrt{\frac{\kappa_3}{\kappa_2}}\, u'_2, \quad u''_3 = \sqrt{\frac{\kappa_3}{\kappa_2}}\, u''_2 .$$

Hence, writing Γ instead of Γ' for simplicity we have:

$$A_3 = -\sqrt{\frac{\kappa_3}{\kappa_2}}\,(u'_2 e^{i\Gamma z} - u''_2 e^{-i\Gamma z}). \tag{11.13}$$

Now assume $A_2 = c, A_3 = 0$ for $z = 0$. It is easily seen that this condition is satisfied by putting $u'_2 = u''_2 = c/2$.

We thus obtain:

$$A_2 = \frac{c}{2}(e^{i\Gamma z} + e^{-i\Gamma z}) = c \cos \Gamma z,$$

$$A_3 = -i\sqrt{\frac{\kappa_3}{\kappa_2}}\,\frac{c}{2}\sin \Gamma z.$$

Hence the occurrence of two waves of frequency ω_1 and ω_2 can generate a wave of frequency ω_3, ω_3 being the sum of the two frequencies ω_1 and ω_2. We thus have the phenomenon of the 'sum-generation' or 'up-conversion'.

Now by the occurrence of the term in η, Γ is very small, so that $\sin \Gamma z$ can be replaced by Γz, and the amplitude $|A_3|$ of the field of frequency ω_3 increases linearly in z, as in the case of the matched waves of frequency 2ω generated by a wave of frequency ω. It is worth remarking that at a distance $\ell = \frac{\pi}{2\Gamma}$ we would have $E_2 = 0$, $|E_3| = \sqrt{\frac{\kappa_3}{\kappa_2}}\,|c|$, so that the whole of frequency ω_2 transforms into the wave of frequency ω_3. Notice that the energy corresponding to the wave of frequency ω_3 is taken for the most part from the wave of frequency ω_1, which has considerable energy. That is, this last wave acts as a pump which, through the occurrence of the wave of frequency ω_2, generates the wave of frequency $\omega_3 = \omega_1 + \omega_2$.

6.12 Let us now examine the case in which this time the system is matched on the wave ω_3, such that:

$$\omega_3 = \omega_1 - \omega_2 \tag{12.1}$$

Again let A_1 be much greater than A_2. Then in computing F(E) we have again to take into account only the terms in $\omega_1, \omega_2, \omega_3$; that is in θ_1, θ_2 and θ_3. However, now we have:

$$\theta_3 = \theta_1 - \theta_2, \quad \theta_1 = \theta_3 + \theta_2, \quad \theta_2 = \theta_1 - \theta_3. \tag{12.2}$$

(11.3) yields, up to terms $O(\eta^2)$:

$$F(E) = 2\alpha(A_1 A_2^* e^{i\theta_3} + A_2 A_3 e^{i\theta_1} + A_1 A_3^* e^{i\theta_2} + c.c.). \tag{12.3}$$

Proceeding as in the preceding case, we have that A_1 can be considered as a real positive constant, while A_2 and A_3 satisfy the equations:

$$\frac{dA_2}{dz} = - i\eta\kappa_2 A_1 A_3^*, \tag{12.4}$$

$$\frac{dA_3}{dz} = - i\eta\kappa_3 A_1 A_2^*. \tag{12.5}$$

Let us now look for a particular solution of (12.4) and (12.5) by putting:

$$A_2 = u_2 e^{\Gamma z}, \quad A_3 = u_3 e^{\Gamma z}, \tag{12.6}$$

Γ being real, with u_2 and u_3 possibly complex.

We then have:

$$\Gamma u_2 = - i\eta\kappa_2 A_1 u_3^*, \tag{12.7}$$

$$\Gamma u_3 = - i\eta\kappa_3 A_1 u_2^*. \tag{12.8}$$

The last equation can be replaced by its complex conjugate, that is by:

$$\Gamma u_3^* = i\eta\kappa_3 A_1 u_2. \tag{12.8'}$$

Whence, multiplying (12.7) by Γ we find:

$$\Gamma^2 = \eta^2 \kappa_2 \kappa_3 |A_1|^2. \tag{12.9}$$

Hence Γ can take only the following values, that are real and of opposite sign:

$$\Gamma' = \eta\sqrt{\kappa_2\kappa_3}\,|A_1|, \qquad \Gamma'' = -\eta\sqrt{\kappa_2\kappa_3}\,|A_1|.$$

Equation (12.8) then yields:

$$u_3' = -\frac{i\eta\kappa_3 A_1 u_2^*}{\eta\sqrt{\kappa_2\kappa_3}\,A_1} = -i\sqrt{\frac{\kappa_3}{\kappa_2}}\,u_2^*, \qquad u_3'' = -u_3'. \tag{12.10}$$

Hence, replacing Γ' by Γ, we find:

$$A_2 = u_2' e^{\Gamma z} + u_2'' e^{-\Gamma z},$$

$$A_3 = -i\sqrt{\frac{\kappa_3}{\kappa_2}}\,(u_2' e^{\Gamma z} - u_2'' e^{-\Gamma z}).$$

Let us now proceed to compute A_2 and A_3 assuming again that $A_2(0) = c$, c being real, and $A_3 = 0$.

These conditions are satisfied if we set:

$$A_2(z) = \frac{c}{2}\,(e^{\Gamma z} + e^{-\Gamma z}), \tag{12.11}$$

$$A_3(z) = -i\sqrt{\frac{\kappa_3}{\kappa_2}}\,(e^{\Gamma z} - e^{-\Gamma z}). \tag{12.12}$$

Hence, when two waves of frequency ω_1 and ω_2 are present, a wave of frequency $\omega_3 = \omega_1 - \omega_2$ can arise. This is the "difference generation or "mixing" phenomenon.

Notice that if Γz is small enough, so that $(\Gamma z)^2$ and higher powers can be neglected, $A_3(z)$ is linear in z as in the former case, while if Γz takes large values $A_2(z)$ and $A_3(z)$ increase exponentially, the larger being the exponent, the larger is $|A_1|$. In other words, we can state that, as in the former case, A_1 acts as a pump that generates the wave of frequency $\omega_3 = \omega_1 - \omega_2$ because of the occurrence of the wave of frequency ω_2.

Bibliography

1. D. Graffi. Su un problema d'induzione magnetica. Rendiconti Accademia delle Scienze di Bologna (11) - I I (1954 - 55), 5 - 12.

2. M. Fabrizio. Problemi di unicità per le equazioni funzionali non lineari del campo elettromagnetico. Nota I e II, Rendiconti Accademia dei Lincei (XLIX), (1970), 268 - 271, 200 - 207.

3. D. Graffi. Il teorema di unicità nella teoria della propagazione del calore con un mezzo di conduttività variabile con la temperatura. Atti Seminario Mat. e Fis. Universita di Modena, II (1948 - 49), 143 - 148.

4. E. Foa'. Sulla transmissione del calore in mezzi isotropi e anisotropi con coefficiente di conduttività variable con la temperatura. Memorie Accademia delle Scienza di Bologna (10) IV (1946 - 47), 119 - 122.

5. J. Serrin. On the uniqueness of compressible fluid motions. Archive for Rational Mechanics and Analysis. III (1959), 271 - 288.

6. J. Serrin. On the stability of viscous fluid motions. Archive for Rational Mechanics and Analysis. III (1959), 1 - 13.

7. D. Graffi. Sul teorema di unicità nella dinamica dei fluidi. Annali di Matematica (4), L (1960), 379 - 387.

8. G. P. Galdi and S. Rionero. A uniqueness theorem for hydrodynamic flows in unbounded domains. Annali di Matematica (4), CVIII (1976), 361 - 366.

9. G. P. Galdi and S. Rionero. On the uniqueness of viscous fluid motions. Archive for Rational Mechanics and Analysis. 62 (1976), 295 - 301.

10. M. De Socio. Influenza di alcuni termini non lineari sul campo elettromagnetico in un gas ionizzato. Rendiconti Accademia dei Lincei, XXXVIII (1965), 860 - 865.

11. D. Graffi. Sul teorema di unicità per le equazioni del campo elettromagnetico in un plasma. Annali di Matematica (L), LXXXIV, (1970), 187 - 199.

12. M.E.Gurtin. The linear theory of elasticity.
Handbuch der Physik, Band VI, 1/2, Springer Verlag (1972), 218.

13. D.Graffi. Onde elettromagnetiche.
Consiglio Nazionale delle Ricerche, Roma, (1976).

14. L.Cesari. A boundary value problem for quasi-linear hyperbolic systems in the Schauder canonical form.
Annali Scuola Normale Superiore di Pisa. (4) I (1974), 311-358.

15. L.Cesari. Sistemi iperbolici e oscillazioni non lineari.
Rendiconti Seminario Mat.e Fis. Univ. Milano XLIV (1974), 139 - 154.

16. L.Cesari. Un problema ai limiti per le equazioni iperboliche quasi lineari nella forma canonica di Schauder.
Rendiconti Accademia dei Lincei LVI (1974), 303 - 307.

17. Y. A. Mitropolsky. Problèmes de la théorie asymptotique des oscillations non stationnaires.
Gauthier-Villars, Paris (1960).

18. P.Bassanini. A non-linear hyperbolic problem arising from a question of non-linear optics.
ZAMP 27, 4, (1976), 409 - 422.